James Crowden is an author a
For twenty years James worke
maker and forester. He is the au
and nonfiction, including *Cider*
Simon Food and Drink Award.

Praise for *Cider Country*:

'James Crowden is Britain's best cider writer. I always turn to his work first when I want top research and inspiring opinions. His new *Cider Country* is the book we have all been waiting for'

Oz Clarke

'Wonderful . . . From the ancient orchards of Kazakhstan to the cider presses of Somerset, fizzing with fruity stories and yeasty historical tales!'

Alice Roberts

'Fascinating . . . Crowden knows this world intimately. He has a gift for evoking the rhythms and smells of cider making'

Henry Jeffreys, *Spectator*

'Crowden writes with an intoxicating lyricism about the great love of his life – cider. Packed with cider-flavoured nuggets of history, magic and folklore, this book will not just make you want to drink the stuff, it will have you packing your bags and moving to the West Country to make it'

Ned Palmer, author of *A Cheesemonger's History of the British Isles*

'James Crowden takes us on the most immersive journey through this drink and tells us the story of the origins of the apple through Kazakhstan . . . It's such a friendly book, a cosy feel . . . This drink fell into decline particularly in the seventies and eighties, and in more recent years we've had this revival of fine cider . . . This book tells that story'

Dan Saladino, BBC Radio 4 *The Food Programme* 'Books of the Year'

'*Cider Country* is a vivid ramble through orchards and history . . . Enormous fun, and effortlessly readable'

Caroline Eden, author of *Red Sands*

'Imagine that Falstaff's got a handful of PhDs, and that he's holding court late at night in a West Country cider house, rumbustiously, outrageously, learnedly, rapturously, fascinatingly. That's Crowden here. Don't miss it'

Charles Foster, author of *Being a Beast*

'A sweeping narrative that never fails to entertain and inform . . . Reading it is like spending and afternoon in the cider barn with your favourite slightly eccentric uncle, the one that has read long and deeply on many subjects, telling stories over an endless stream of pints. What a delicious way to spend one's time'

Cidercraft

ALSO BY JAMES CROWDEN

Cider: The Forgotten Miracle
Ciderland
The Frozen River

Cider Country

How an Ancient Craft Became a Way of Life

James Crowden

WILLIAM
COLLINS

William Collins
An imprint of HarperCollins*Publishers*
1 London Bridge Street
London SE1 9GF

WilliamCollinsBooks.com

HarperCollins*Publishers*
1st Floor, Watermarque Building, Ringsend Road
Dublin 4, Ireland

First published in Great Britain in 2021 by William Collins
This William Collins paperback edition published in 2022

1

A catalogue record for this book is available from the British Library

ISBN 978-0-00-839357-1

Typeset in ITC Garamond LT by Palimpsest Book Production Ltd,
Falkirk, Stirlingshire

Printed and bound in the UK using 100% renewable electricity at
CPI Group (UK) Ltd

This book is produced from independently certified FSC™ paper
to ensure responsible forest management.

For more information visit: www.harpercollins.co.uk/green

To the wild apple forests of Kazakhstan
& the traditional Cider Makers of England

'Stay me with flagons, comfort me with apples'

Song of Solomon 2:5

'Cider, generous, strong, sufficiently heady, excites and cleanses the stomach, strengthens the digestion and infallibly frees the kidney and bladder from breeding the gravel stone . . .'

From John Evelyn's *Pomona*, Royal Society, 1664

'Constant use of this liquor . . . hath been found by long experience to avail much to health and long life; preserving the Drinker's of it in their full strength and vigour even to very old age'

John Worlidge, *Vinetum Britannicum –*
A Treatise on Cider, 1676

'Bread is the Staff of Life – but Cider is Life Itself'

Horace Lancaster, Devon cidermaker, 1966

CONTENTS

Introduction	1
1. Wild Apple Forests	11
2. Apples in Antiquity	31
3. Apples of the Imagination	61
4. Medieval Cider	88
5. Orchard Renaissance	110
6. Maritime Cider	136
7. Sparkling Cider	164
8. Georgian Cider	231
9. Twentieth Century Cider	299
10. Twenty-First Century Cider	346
Epilogue	409
Acknowledgements	414
Bibliography	417
List of Images	424
Index	429

Cider weights and measures

Quart – 2 pints
Firkin – 2 quarts
Gallon – 8 pints or 4 quarts
Six – 6 gallons
Kilderkin – 18 gallons
Barrel – 36 gallons
Hogshead – 54 gallons
Puncheon – 84 gallons
Butt – 108 gallons
Pipe – 120 gallons
Tun – 2 pipes or 240 gallons

Introduction

THE CIDER FARM

The autumn clock is ticking. A steady stream of tractors and trailers bring an endless river of mellow fruit into the cider farm, gathered in from orchards across the county. Mountains of apples lie piled up in heaps outside the barns, ready for pressing. It's important not to press them too soon. You need to let their flavours mature upon the ground in the soft autumn sun and reach their own inner perfection. This is one of many secrets.

Cider is a way of life that you fall into by accident or chance. A secret world that beckons you from afar. Just like falling through a trapdoor into another dimension. But even finding a cider farm can be an adventure in itself. You sometimes find yourself asking for directions at pubs or peering over hedgerows into the middle distance just to get your bearings. Sat nav rarely works down these narrow, winding lanes, so intuition and dead reckoning are better comrades. It is as if we are mysteriously drawn by an invisible force of nature, or maybe we are genetically programmed to make bee lines for these farms, which every spring are surrounded by an undulating sea of pink and white blossom that runs for miles.

Fig. 1.1 'Cidermaking in the 17th century was a gentlemanly art', by John Worlidge, 1678

Time on these remote cider farms seems to stand still for most of the year. Upon arrival your mind also slows down. You sniff the air and stroll about the yard. And then, after a few minutes, your senses become attuned to this more relaxed pace of life. You start to notice small things.

To get a real flavour of this 'forgotten' world you simply ring the bell, climb a few steps and enter the large, creaking doorway that leads into a musty barn full of wooden barrels and cobwebs, hogsheads and pipes. You walk in a little further as if entering a cavern, and in the dark barn squeeze past enormous vats. Here you find the cider press: a heavy-duty hydraulic piece of kit that exerts great pressure when you want it to. In the nineteenth century this key invention took over from the great wooden presses with screw threads that dated back to medieval times. But today your life is dictated by the speed of pressing. The chopped apple is built up into a square with layers about three feet high and held in place with slatted ash boards and large brown hessian cloths. This is called a 'cheese'. In the old days they would have been made with long straw, which was folded in neatly to keep the apple intact when under pressure.

There is a rich aroma of crushed apple hanging in the air, like an exotic perfume, so rich that you can almost taste it. If you listen carefully you can hear the sound of cider bubbling away, fermenting in the background. Nature doing its thing.

There are only two of you building the cheeses, one on either side, and after every cheese you swap sides. One controls the apple chute from above and the other positions the boards and cloths. Like most manual work, pressing cider has a rhythm all of its own. There is always the gush of running water as apples are hosed down into the apple pit. From here the clunking elevator takes the apples up to the loft where the mill chops them up with a high-speed whirring noise.

Each layer of apple must be even. You need a good eye and an even touch as you spread the pulped apples with your hands. Once built, you push the cheese into the press and start the pump. The apple juice pours out in an orange-brown waterfall that is piped away. The dry, spent pomace is then thrown into a muck spreader and taken out to the fields for the cattle. Two presses work simultaneously and can produce 100 gallons every twenty minutes. This is where it all happens.

Slowly the smell of apples works its way into your mind. You live and breathe cider apples every day. Your hands are stained black with tannin. And there is no way you can wash it off. Just like the smell of apples, it is deeply ingrained. You wear a hat, a smock and an apron to keep the juice from running all over you. It is like being in the engine room of a small tramp steamer; a wonderful way to keep fit and warm. Some tall oak vats are over 150 years old and are so vast it takes a week to fill them.

And then it's time for the ritual tastings. Last year's cider. Communion with the apple god. Reward for your hard work the previous autumn. You consult the barrels and sample the cider, accompanied by the squeak of a wooden tap as it is turned on and then off. First one barrel, then another. The golden nectar in the glass is held up to the light. Sampling is an essential part of the process. A way of getting your bearings and your taste buds attuned. Rich, mellow tannins, some acidity, fruit on the nose, strong, golden colour and a refreshingly complex depth that goes on and on. If you are very lucky you may get to taste some sparkling bottle-fermented cider, the *pièce de résistance*.

Once the cider has made a favourable impression upon your palate, you are usually hooked for life, in the most pleasant way possible. Your soul becomes apple-shaped.

Many cider enthusiasts return again and again to get their flagons filled. But I went one step further and dived into the deep end, working for twelve seasons on Burrow Hill Cider farm in Somerset. It was the best way to get a real feel for the cider world. You learn by osmosis.

* * *

Cider is the gateway into a rustic way of life, which lies buried deep within our own consciousness and ancestry. An antidote to the modern, fast conveyor-belt, computer-driven life which hermetically seals people inside offices and isolates them from the land and each other. The cider farm is at the centre of a vast network of orchards. And if you work there you are at the very hub of it all, pressing apples eight hours a day in a large barn, where whole orchards are pulped and crushed. This is the very heart of cider country.

To many people cider is often just a drink with romantic notions of a fine hot summer's day, watching cricket or tennis, going for a swim in languid rivers or having a picnic in a field. But cider is far more than just another drink. It has invisible connections and an ancient pedigree with a remarkable story. Beneath the surface there are thousands of years of untapped history and apple archaeology, Greek legends and Roman apple goddesses, apple dining clubs, Norse sagas, Irish myths, monastic orchards, herbalists and civil wars, political ferment, medicine, scientific experiments, sea voyages and apple migrations. Cider riots and cider glasses, cider poets and inventors, pioneers and excisemen, mother orchards and apple days, cider laboratories and cider clubs. Cider is a law unto itself. And so are cider makers.

Even the word cider has an interesting pedigree. From the Biblical Hebrew *shekar*, via ancient Greek translations, it comes down to us as *sicer, sicera, sidra* or *sikera*, which sounds a bit like a Greek island where orchards and fair nymphs dart between apple trees. *Shekar* meant strong drink and may have been fortified with honey, tasting rather like sweet Samos wine. Old French *sidre* became *cidre* and with Norman French after 1066 it became cyder, and now *cider*. Both are used today.

* * *

This is autumn. Not just on this cider farm but on cider farms all across cider country in the South West. Everything speeds up as cider apples are picked in the orchards, either by machine or by men and women on their hands and knees. It is hard work in all weathers. Each cider farm has its own unwritten rules, cadences and codes of behaviour. Its own morals, culture, language and ethics: its own *cider philosophy*.

This was how I heard all the cider apple names and their stories. Old grafters in battered tweed jackets and binder twine gave you the benefit of their wisdom. Dabinett and Stoke Red, Kingston Black, Yarlington Mill, Tremlett's Bitter, Harry Master's Jersey, Foxwhelp, Brown Snout, Ellis Bitter, Slack ma Girdle, Tom Putt and Pig's Nose. Just to name a few.

Each cider apple has its own devoted following and stories to reflect the names of the farmers who discovered them or places where they were found. These farmers worshipped their cider apples as if they were patron saints. Their orchards were their temples. True believers keen to engage in conversation and convert the uninitiated.

As for cider, it always comes down to taste. Farmhouse cider made from cider apples is a very particular drink. Subtle and complex. Aged in oak. Totally and utterly different to beer. Cider is the unsung hero of the English landscape and, for the cognoscenti, cider is a *way of life*. A history which defines cider and makes it an English gem.

Real artisan cider is made from high-tannin cider apples with prolonged depth of flavour, a rich world ripe for exploration. Their taste is vital. But what happens when you bite into a cider apple? What are your first impressions? Why do they taste so different to dessert or eating apples or cooking apples? What is it about them that makes them special?

Some apples are sweet, some sharp and acid, others bittersweet or bittersharp. Many of the best cider apples have high levels of tannins that give a strong astringency. These tannins give the cider the deep, complex flavours that the cider maker is looking for. It is a hidden language of taste and intuition. The golden measure, the holy grail. So very different from the supermarket apples which have often travelled halfway round the world in a temperature-controlled, chilled container. Chalk and cheese.

There are hundreds of cider apples out there, if not

thousands. It is the inherited DNA, the complex inner structure that gives the variant aroma its identity. All this history has to be considered in a few seconds of tasting.

Sometimes it takes years of experiments to understand a single apple. The fruits of one's labours, blended, bottled up, matured and then drunk. Wild and cultured yeasts also play a crucial part in fermentation. A process as old as the hills. Sometimes wild cider apples can be found growing in the hedgerows and that is when cider becomes very interesting. New varieties can be named and propagated.

* * *

I developed an interest in wild apples while working at Burrow Hill in the mid-1990s. Word got around that apples didn't originally come from Dorset, Somerset or even Normandy, but a place called Kazakhstan which had been under the Soviet thumb. An American journalist called Frank Browning pitched up at Burrow Hill Cider Farm and told us all about it. He had grown up on an apple farm in Kentucky, so he knew what he was talking about. He had been out to Kazakhstan with some American scientists from Cornell University who had taken genetic samples to help with their experiments. He said these wild apple forests were in the Tien Shan – the Heavenly Mountains in eastern Kazakhstan. These forests were full of bears and were reckoned to be the origin of all known apples in the world.

My ears pricked up. I was fascinated by Central Asia and had been to Afghanistan in the 1970s. I had seen wild grapes growing in Kurdistan, wild apples growing in the Hindu Kush and wild apricots in the mountains of Ladakh, so I knew it was perfectly possible. But the idea that apples had migrated all the way from wild apple forests in Central Asia to France and then to the cider farms of Somerset and Herefordshire was quite extraordinary. I realized then that the cider world was much larger and more expansive than I thought. Mind-blowing.

Almost everyone has an affinity for apples and orchards when they are growing up, and yet how many people really know how cider is made? What exactly is our relationship to this complex and intriguing drink?

I began to dig a bit deeper and investigate the history of cider making, a fascinating journey that has taken me into archives and libraries, museums, secret valleys, remote farms, leather-bound books and darkened barns. But one question always dogged me. Where did the cider apples that we pressed every year with high tannins originally come from, and how did they get here? Did the apples come all the way from Kazakhstan or did they slowly emerge by crossing with our indigenous small, bitter little crab apples? Who discovered them? Or did the cider apples discover us?

1

WILD APPLE FORESTS

The road to Kazakhstan was not just a physical and spiritual journey. It was an intellectual and scientific adventure as well as a voyage of discovery into the mysteries of apple genetics. Central Asia held the key.

For several thousands of years no one really knew the exact location of these wild apple forests. Were they mythical, living in people's imaginations as a strange, otherworldly Garden of Eden, or did they actually exist? Most people did not give it a second thought. Down on the cider farm, apples lived in orchards and hedgerows and could be scrumped. That was it. And if apples came from anywhere, it was either from God or somewhere further east.

Wild apple stocks are notoriously elusive. They often survive in remote valleys and rugged mountains where they can flourish. But apples have one distinct advantage over some other fruits. They can spread of their own accord, either with the help of birds, bears, bees, camels, and horses, or with the help of nomads and early farmers.

Nomadic trade carried apples far and wide. Fruit was always a valuable commodity.

Early archaeological remains of apples point to the Middle East and Caucasus. Carbonized fruits, including crab apples from 6500 BC, have been found at Çatalhöyük in modern day Turkey. Çatalhöyük is one of the earliest excavated towns in the Middle East. Here, evidence has been found for peas, crab apples, juniper berries, acorns, almonds and grapes. An ordered, even egalitarian existence.

Even as far away as China there are rumblings about apple culture. In 5000 BC Chinese diplomat Feng Li reputedly gave up his office job to graft apples, pears, peaches and persimmons - perhaps that's why they call it gardening leave.

Apples also crop up in the Sumerian city of Ur of the Chaldees c. 2500 BC, where eleven charred apple rings threaded on a string were found in a cemetery by archaeologist Leonard Woolley. Queen Puabi was buried here with all her jewellery and 74 attendants. 6 men and 68 women all decked out in fine jewellery before they were killed.

The Sumerians were highly sophisticated, with an urban lifestyle and well-organized agriculture. Their apples would have undoubtedly come from further north, in what is now Kurdistan. Apples also feature in Queen Puabi's diadem, which shows stylized gold apples and dates along with bulls, stags, gazelles and rams. Some might say they were pomegranates, but pomegranates do not grow as clusters. Apples often do but not always. There is a later Sumerian hymn to Inanna in praise of the goddess that says, 'You are she who creates apples in their clusters.' This may well have been written by the world's earliest known poet, Enheduanna, princess and High Priestess of the goddess Inanna.

Other empires have apple remains and apple-related texts: the Hittites, the Urartu and the Armenians, whose cultures all favour the apple in some way. The Armenians also worship Mount Ararat, famous for apricots and apricot brandy. As always it is the early shifts towards settled agriculture that eventually leads to the planting of orchards. In dry climates early apple fragments can be found as well as written comments which refer to gifts and feasts, orchards and even fines for damaging orchards or stealing fruit. Farming has its advantages as well as its disadvantages. If you have an orchard you must tend it and protect it and harvest it. That way the fruit has a high economic and spiritual value.

In northern Mesopotamia a tablet from about 1500 BC records that a man called Tupktilla of Nuzi, an Assyrian city, sold an orchard he had inherited for three sheep. The Hittites had fines for people causing wanton damage by fire to orchards. Stiff fines for each apple tree, vine or pear tree or pomegranate. Six shekels of silver and replant the plantation. One Hittite orchard had 40 trees, and a royal palace in Nineveh had an orchard with 42 different types of apple and gum tree. Apples and orchards were in the ascendancy and this was 1000 BC. The Garden of Eden in Mesopotamia had a very real identity.

The Persians were also very keen on orchards, they conquered vast tracts of the Middle East and even threatened Greece. Military conquest was flavour of the month. One of the youngest Greek generals was the Athenian

Xenophon (431–354 BC). He was a philosopher and student of Socrates. Very importantly he describes a Persian walled garden or *pairidaēza* that he visited. Xenophon called the orchard *paradeisos*, which describes not just the wall and garden but the enclosed parks that Persian nobles loved to hunt in. Upon his return to Greece, Xenophon made a walled garden and orchard for himself. That is how the concept of Paradise and the word *paradeisos*, slipped first into the Greek world, then into the Roman world and then into our own.

But the true source of apples was still an enigma for Greek philosophers and geographers. Another philosopher general, born just before Xenophon died, was Alexander the Great, (356–323 BC). He also liked apples and on one of his campaigns in Central Asia, got quite close to Kazakhstan. Alexander sent apple grafts and saplings back to Athens to his old philosophy teacher, Aristotle, who had taught him when he was a boy. Apples and philosophy were often joined at the hip. Maybe they always have been.

* * *

But where exactly is Kazakhstan? And what secrets do the wild apples forests actually hold? Kazakhstan, 'The Land of Wanderers', is still semi-nomadic. Some Kazakhs still live in yurts. Horses, bears, camels, sheep and snow leopards abound. Kazakhstan is 4,000 miles from England. If a horse caravan travels at 12 miles a day, that is a year's continuous walking with the odd rest day thrown in. These days the wild apple forests are a happy hunting ground for pomologists and plant biologists.

Kazakhstan is vast. Larger than Algeria, Congo, Mexico, Greenland and Saudi Arabia. Ten times the size of the United Kingdom. In the northeast is Semipalatinsk where the Soviets tested atomic bombs. It was also where Dostoyevsky went after four years' exile in Siberia. He rather liked it. Pine forests, lakes and clean air. He married his first wife Maria there.

But it is in the southeast corner of Kazakhstan, close to Almaty and in the steep foothills of Trans-Ili Alatau mountain range and the Tien Shan, that the apple forests lie. Apple forests that until the 1950s were very extensive. Many were destroyed to make way for wheatfields and housing. What is extraordinary is that nearly all apples in the world have come from this one region. By a quirk of climate and geography they escaped the punishment of the Ice Ages and continued on their journey of diversification. Thank heavens they did. Otherwise we would not have any apples or other similar fruit today.

But how did that happen? And who first realized that wild apples lived there? This question has had scientists perplexed for over 200 years because apple genetics – like apple myths – are notoriously difficult to disentangle and pin down.

Genetics are very interesting. Everybody should have a few wild ancestors in their background. By surviving in the wild, apples become stronger, more diverse and able to resist disease and variations of climate. Today there are more than 7,500 cultivars of apples throughout the world

- the National Fruit Collection at Brogdale in Kent has over 2,300 examples. Most English cider counties now have mother orchards to preserve their own varieties for future generations.

Rather like us, apples picked up wild genes along the way. Tamed, then planted out in walled gardens and orchards and given English names. Some dessert apples and pears still retain French names. So what we often think of as quintessentially English apples are not originally from England at all. Their ancestors were out East. But a long time ago.

The discovery of the wild apple forests in Central Asia by early scientists is as fascinating as the apples themselves, but the intellectual thread was nearly lost several times. The wild apple, *Malus sieversii*, is native to the mountains of Central Asia in southern Kazakhstan. It is named after Johan Sievers, a German medical botanist who, while looking for wild rhubarb, stumbled across the apple forest in 1790s. This is the Tien Shan, the Heavenly Mountains, north and west of the Taklamakan Desert and directly north of the Tarim Basin. Its capital, Almaty or Alma Ata, means 'apple mountain' in Kazakh or 'father of apples' in Russian.

When Johan Sievers got back to St Petersburg he died suddenly aged only 33. Luckily his observations were published in journals. Forty years later a German-Estonian botanist, Carl Friedrich von Ledebour, made another trip east. Lebedour gave full descriptions of these wild apples and named them *Pyrus sieversii*, a tribute to the earlier work of Johan Sievers. So far so good.

Today Kazakhstan shares its borders with Russia, China, Kyrgyzstan, Uzbekistan, Turkmenistan and the Caspian Sea.

The word Kazakh comes from ancient Turkic word *qaz*, 'to wander', reflecting the Kazakhs' nomadic culture. 'Stan' means 'land' in Persian, as in Tajikistan, Uzbekistan and Afghanistan. So the wild apple forest is situated in the Mountains of Heaven in the 'land of wanderers'. Apples are also nomadic.

When they came west to the Middle East and Europe these wild apples from Almaty mingled genetically with indigenous little crabs, *Malus sylvestris*, the 'woodland apples' dotted all over Europe and the Middle East. These woodland apples are often tucked away in hedgerows and forests, bitter little numbers that somehow survived the Ice Age. Crab apples were used for crabbing the parson. Cattle love them. They are more important than you might think.

Nikolai Vavilov

As if in a fairy tale, the wild apple story lay dormant for nearly one hundred years until the son of a Moscow textile factory manager once again stumbled across the wild apple forests. The intrepid traveller was the plant geneticist Nikolai Vavilov (1887–1943), a larger-than-life character brimming with charm, fluent in English, French and German. He spoke Spanish and Italian and . . . the language of apples. The Russian geobotanist Leonid Efimovich Rodin (1907–1990) once said: 'A single chat with Vavilov was enough to charge one's batteries for a whole month.'

Just before the First World War Vavilov visited Europe and studied plant immunity with the British biologist

William Bateson (1861–1926) in Cambridge. Bateson helped establish the science of genetics, and first suggested using the word 'genetics' (from the Greek gennō, γεννώ, 'to give birth') to describe the study of inheritance and the science of variation.

Bateson also first used the term 'genetics' publicly in London in 1906 at the Third International Conference on Plant Hybridization. Three years later Wilhelm Johannsen used the word 'gene', cutting-edge stuff in botanical circles, but even Bateson had an uphill battle to get his peers to accept his theories.

The bilingual Vavilov was Professor of Agronomy at Saratov University on the Volga. In 1924 he became director of the Lenin Academy of Agricultural Sciences at Leningrad. He travelled very widely – Central Asia, Persia, Afghanistan, Europe, Italy, Spain, Peru, Guatemala, Canada and the USA, even to the headwaters of the Nile – always searching for wild grains – wheat, barley, maize, apples and apricots, anything he could get his hands on. Samples were sent back to his wife and stored in Leningrad. He had a truly global view of plant genetics. Since childhood, Vavilov had been obsessed with trying to alleviate famine. He'd seen starvation first-hand as a boy in Russia. That is what drove him to find new crops.

I first heard Vavilov's name from my great-uncle, Professor Kenneth de Burgh Codrington, who had been in Afghanistan as an archaeologist in the 1940s, collecting wheat samples for Cambridge. Dr Schuyler Jones of the Pitt Rivers Museum in Oxford also mentioned Vavilov in passing. Vavilov had travelled to Afghanistan in 1924 and had crossed the Hindu Kush on foot, visiting Badakhshan and Nuristan.

As for wild apples, Vavilov did not reach Almaty until 1929. He was only there for a day or two but what he saw bowled him over, and changed his fundamental understanding of apple genetics. Vavilov describes the moment in his book *Five Continents*:

> Thickets of wild apples stretch out through an extensive area around the city and along the slopes of the mountains, here and there forming real forests. In contrast to the very small wild apples in the Caucasian mountains, the Kazakh wild apples have large fruits and they don't vary very much from cultivated varieties . . . We could see with our own eyes that this beautiful site was the origin of the cultivated apple where cultivated apples did not rank markedly above wild ones and where it was difficult to distinguish between them. Some of the forms in this forest were so good in respect of quality and dimensions that they could be grown directly in a garden . . .

In a sense Almaty was not just the birthplace of apples but the birthplace of apple myths, legends that were eventually grafted onto Greek philosophy and Roman horticulture. What is also fascinating is that Vavilov's hunch about Almaty being the true home of wild apples has been proved beyond doubt by modern-day geneticists.

Philip Forsline and other US scientists from Cornell were there in the 1990s. Subsequently Dr Barrie Juniper from Oxford University visited the Tien Shan several times and wrote a book about it. Then, in 2012, a young French geneticist called Amandine Cornille and her team pulled out all the stops. After years of work they cracked the complex European apple code and finally unravelled the apple's secret enigma and its links to crab apples.

Cornille had studied in France and Sweden and is now based in Zurich as a researcher at the Swiss Federal Institute of Technology. With the help of many assistants, she undertook a vast sampling of apples from the Basque region in Spain right through Europe, Central Asia and China. They proved that the mountains around Almaty were indeed the source of all known domestic apples. Links with European and Siberian crab apples were more significant than at first thought. Gene traffic was not just a simple route westward, but a complex sharing of genes backwards and forwards along the whole apple corridor.

What is also very interesting is that some of the wild apples in the Tien Shan in Kazakhstan turn out to have high levels of tannin before they ever start their journey west, roughly seven times the tannin of normal domesticated varieties. The key to good cider, as always, was tannin.

But what of Professor Nikolai Vavilov, the agronomist and geneticist? Why do we not hear more about him? His is a very sad story indeed, a Russian own goal of mammoth proportions where communism failed its most intelligent scientists.

In 1929, the year Vavilov reached Almaty, Stalin started his ill-fated scheme for agricultural collectivization which ultimately led to Vavilov's demise. Stalin was obsessed with Trofim Lysenko's outdated and fatally flawed plant-breeding ideas. He therefore purged all scientists who dared to contradict Lysenko. Many Russians starved because of Lysenko's crop experiments.

Lysenko's approach to agriculture led to terrible crop failures and, like many Russian scientists, Vavilov openly criticized his methods. It cost him his life. In 1940 Vavilov was arrested and sentenced to death, which was later commuted to twenty years in prison in Saratov, the very town where he had once been professor. Vavilov starved to death in January 1943. Not many food scientists die of starvation. Unknown to Vavilov, the Royal Society in London made him a fellow just before he died, a rare honour for a Soviet scientist. When he died Vavilov was 55. He had wanted better crops to feed Russia in time of crisis and war. Stalin had scored an own goal.

What is also remarkable is the tenacity of Vavilov's assistants, during the siege of Leningrad, who protected the basement in which his precious seed samples, covering a vast range of crops and fruits, were secretly stored. There were 250,000 samples in all. The siege lasted 872 days. The loyal scientists guarding the seedbank refused to eat its contents, even though by the end of the siege nine of them had died of starvation.

One of Vavilov's students back in Almaty who survived the purges and the Second World War was a young Kazakh called Aimak Dzangaliev. He took on Vavilov's work with apples and spent most of his life saving wild apple forests. Like many scientists, Dzangaliev was sacked by the authorities and downgraded from his job because of his interest in wild apples. His sister was tortured and his brother-in-law disappeared. More than 3,000 Soviet scientists were fired, imprisoned or executed for attempting to oppose Lysenko and his ideas on genetics. Science was very cruel, truth – a rare beast.

For most of his life Dzangaliev kept his head down and worked alone. Then in the 1970s the physicist Andrei Sakharov highlighted the work of Vavilov and his fate became known outside Russia. In the 1990s Dzangaliev was reinstated and in the process his lifetime's work was recognized. Frank Browning, the Kentucky apple farmer, was the first western journalist to meet and interview Dzangaliev. Browning mentions all this in his excellent book *Apples*, published in 1998.The world was now at long last interested in Vavilov, the Tien Shan and wild apples.

The moment of truth

Apple DNA was notoriously difficult to disentangle even though some of the best brains in the apple world were working on the subject. Kazakhstan was also very far away. Occasionally I saw odd articles in journals about apples in the Tien Shan. I went to one of Barrie Juniper's lectures on apples which was fascinating, all about bears and horses, nomads and the Silk Road, and even bought his book. But I was still curious about the relationship of Kazakhstan to cider and cider apples.

And that was why I was keen to talk to the biochemist Andrew Lea who had worked on cider tannins in the 1970s and 1980s at Long Ashton Research Station outside Bristol. We had judged cider together on several occasions in Hereford and Somerset and often agreed on what was a good or even exceptional cider. We also chatted fairly often over the phone about cider history.

Then, out of the blue in 2017, I discovered that Andrew had just been to Kazakhstan to see the wild apple forests for himself. More importantly, Andrew wanted to taste the wild Kazakh apples and explore their hidden depths, to see if cider apples actually grew in the wild. Or, to be more precise, would the wild apples of the Tien Shan make good cider?

Andrew was accompanied to Kazakhstan by three other cider makers, and they went out there to answer that very question. A bold move. I wanted to find out more. Andrew Lea had also been to one of Barrie Juniper's apple talks but, having been involved with cider for most of his life, Andrew looked at the wild apple forests from a slightly different angle. He found the talk riveting but realized that there were some key questions still to be answered. There was little, if any, mention of tannin and cider apples. Vavilov and Dzangaliev were not, I suspect, hardcore cider drinkers.

This set Andrew thinking. Was it possible that wild apples with strong tannins existed in this forest? And, if so, would these wild apples make good cider? An intriguing hypothesis which had to be tested.

Up till this point, everybody had assumed that sweet dessert apples travelling west had genetically crossed with wild crab apples along the way, and that these were the ancestors of the astringent cider apples so desired by cider makers. Apparently no one had asked the obvious question: do cider apples with high tannins exist in the wild in the apple's own backyard? The only way to answer that question was for someone to go to Kazakhstan in the late summer when the wild apples ripened. When Andrew's retirement beckoned, the coast was clear. An apple pilgrimage was on the cards.

Andrew takes up the tale with gusto. 'Fast-forward to 2017 and a chap living in Kazakhstan called Alex Thomas contacted me. He had the idea of making cider from local apples and to develop a local cider culture in Kazakhstan. No cider there as yet. Alex needed to explore the apple forest and thought it would be nice to bring along some of his new-found cider friends.' Andrew Lea and three other cider makers: an American, a French-Canadian and a food chemist from Gloucestershire. A good team.

In August 2017 they flew into Almaty and visited two locations: the Ile-Alatau National Park and the Dzungarian Alatau, right on the Chinese border. They were escorted everywhere. 'We looked at wild apples very carefully. Every tree is different, but our real concern was: could we find within this gamut of flavours in Kazakhstan any that we could call cider apples?

'By the end of the first afternoon we had answered that question. The answer was "Yes". Emphatically YES! We were all agreed on that one. A result within twenty-four hours.

'Many of the wild trees had the characteristics of cider apples: tannin, acidity and sugars. High tannin, hard tannin, soft tannin, some had high acid, some were aromatic. They would have made good cider if people had bothered. Nobody had looked at that.

'Then we went to the botanic gardens where Dzangaliev's apple collection is. Over the years he brought back about sixty different types of apples. These were planted in the Almaty botanical gardens as source material for scientific measurements. There is even one row labelled CIDER but there was no evidence of cider making. In the institute they measured all the parameters of the apples and logged them religiously. All the chemical data of single varieties. Vavilov would have been impressed.'

Garden of Eden?

What interested me was not just the genetic patterns that had evolved of their own accord, but the discovery by Andrew Lea and his friends of a fully laden larder of tannin-rich cider apples right there in the middle of wild apple forests. Just where nature had put them. The road to Kazakhstan was the road to a cider apple Damascus. Here you had it all. Apart from apples there are pears, apricots, plums, walnuts, bears, almonds, mulberries, rhubarb, even hops. All growing wild.

This dry and landlocked region was one of the few places spared from snowsheets during the last Ice Age, which meant that the wild fruit trees had not been destroyed. The survival of these forests was a miracle that was deeply embedded in storytelling, mythology, religion and now science. Almaty was the gateway to the wild apple forests – the original Garden of Eden?

On the second day they all went northeast to the

Dzungarian Alatau and passed through Sarkand on the way. At Lepsinsk forestry camp, they saw a very large apple tree, their pride and joy, the largest wild apple tree recorded in the world. The trunk was 3 metres in circumference, 17 metres high and about 300 years old. About the same size as old perry pear trees in Herefordshire. And it was very healthy. 'The apples were alas inedible.'

'We discussed propagation. Bears in the forests and horses along the trade routes. In Topolevka, we met Professor Isin. It was here they served up a traditional meal in a yurt and as I was the oldest in the group I was offered a sheep's head. I had to ceremonially cut the ear off.

'Professor Isin works in the botanical institute in Almaty. He had worked with Dzangaliev. Quite a talk we had. He said that one of the most important things was that when apples were eaten by bears, apple pips had to pass through the animal's gut to get the germination going.' An apple rite of passage. Maybe that is why bears and apples often appear in stories together. 'The wild apples were really quite large, much larger than most dessert fruit.'

The curious thing was that when Andrew was at Long Ashton he never once had a discussion with his colleagues about where cider apples came from. 'We just assumed they were already here. Every culture in Europe and Western Asia always said they came from further east. The Greeks got them from the Persians, the Persians probably got them from Turkmenistan which borders Kazakhstan.' Trade was rife. But no one realized quite how far they had actually travelled.

For Andrew Lea the whole trip was fascinating and he felt very privileged, part of an historic moment proving the cider apple link to Kazakh wild apple forests. Tannins all the way. The secret was simply to be in the right place at the right time of year to test their hypothesis with four expert witnesses.

One outstanding question was, 'Why was no cider made in Kazakhstan?' Another was, 'Why was there no cider culture?' Especially when the raw material was all around them. Andrew suspected it was to do with the nomadic lifestyle. What the Kazakhs did do was gather apples, dry them and carry them on their journeys. To rehydrate? Simply boil them up and drink the juice. As nomads they were never in one place long enough to make cider. Fermentation takes a while and storage would be a problem. You need barns, barrels and cider presses.

The Silk Road had its own culture and nomadic imperative. Cider making would simply not have fitted into their annual cycle. It is lucky that Vavilov realized the true importance of the wild apple forest at first glance and that Aimak Dzangaliev kept the flame of apple research alive.

So these wild apple forests, when at their fullest extent, reached into Xinjiang and China as well, and must have appeared like a Garden of Eden. The gardeners never toiled, and fruit appeared as if by magic every year.

Apple enlightenment

One thing I wanted to know from Andrew was what it felt like to be in these wild apple forests which were the source of all known apples. Andrew remembers a very strong emotional feeling which he found hard to express: It was 'a kind of pilgrimage to visit the apple Mecca, the apple Holy Land' . . . 'Whatever people do on pilgrimages it was a similar sort of thing – we all took our first bites of the apples and looked around at the landscape. It was almost a religious experience.'

To Andrew one of the sad things was the obvious depletion of the forest. Stalin had chopped down all the apple trees and ploughed up the steppe. Land that grows good wheat can also grow good apples. The soil requirements are the same. Stalin was trying to turn Kazakhstan into the breadbasket of Russia to send grain around the rest of the Soviet empire.

'Occasionally you get single apple trees that have sprung up all on their own on field margins and they are the relics, survivors, hangers-on. We were told that if one of these seedlings appears, by law you are not allowed to cut it down and must allow it to develop. The Kazakhs are very embarrassed by what happened in the 1940s. But it wasn't their fault.

'In the old days, there were thousands of acres of apple forests on these slightly undulating flat lands. The only forests that remain are on the hilly slopes that couldn't be cultivated. That is terribly sad. The forest used to come right round into the city. No trace of that now.

'As for the Kazakh apples they were exactly what we wanted for cider making. Good tannins, body and the right chemical structure.' So he had no doubt about the genes.

Genetically there were 'lots of technical issues as to how much *Malus sylvestris* has contributed to the domestic apple in Europe as it came along the Silk Road. All I can say is that any of those apples could have been the direct progenitors of the cider apples we have today.'

So they were looking at the apple forest with the same excitement as Vavilov had done in 1929, but with a cider apple perspective. A historic moment. 'Kazakh cider apples would make a very good blend. The wide range of apples all over the world is fully expressed in those wild forests.'

The legacy of Nikolai Vavilov and the other botanists is extraordinary. Wild ancestors indeed. And we are the beneficiaries of their tenacity and keen observations. What is also under threat is the apple forest itself. As noted by the IUCN (International Union for Conservation of Nature) Red List, the habitat of the apple forest has declined by over 70 per cent in the last thirty years. Its status was already 'Vulnerable' in 2007. Mining and agriculture have encroached on its habitat. Its gene bank has such diversity that it is of vast importance and irreplaceable. It is also a key part of our history and identity. A key ingredient in our myths, legends and philosophy. Vital even to our health and sense of wellbeing.

In a sense all our orchards and walled gardens are a mirror image in miniature of these wild apple forests. Mythology was correct all along.

Cider, and our *Cider Country*, has its roots in the Heavenly Mountains – the Tien Shan – Paradise indeed.

So when you next bite into an apple think about Kazakhstan, wild apple forests, nomads, bears, camels and horses. And, of course, of Nikolai Ivanovich Vavilov and Aimakh Dzangaliev.

2

APPLES IN ANTIQUITY

Apples were very popular in the classical world but how did they arrive in Greece? The wild apple forests are nearly 3,000 miles east of Athens. The answer may well lie in the Greek trade routes. They set up trading colonies not just in the Mediterranean but all along the Black Sea coast. Some of these colonies traded with nomadic tribes like the Scythians, who had links to Central Asia. Scythians were mentioned by Herodotus and were admired for their culture and skilled goldwork. They even had golden apples. It was with the Scythians that Alexander the Great picked a fight on the borders of Kazakhstan. This was the Battle of Jaxartes, just south of Tashkent. Alexander was only 500 miles from Almaty. So near and yet so far.

Another more southerly trade route led through Persia, Armenia and Turkey. Apples travelled well and grafting was a neat way of propagating fruit. Each town and oasis had its own orchards. What is known is that apples have been cultivated in Greece since the eighth century BC.

Greek philosophers were always hot on the apple trail. They turned contemplating the apple into an intellectual activity and philosophical pursuit which exercised their minds.

At first, apples were rare, exotic and expensive, hence notions of earthly Paradise. They had value, both economic and as fruits of pleasure. Eating or even stealing apples was a metaphor for life. Apples were easily transported and given as a luxury, a timely gift, reciprocity and delight, a rich flavour imparted. Love tokens. Apples lent themselves to the mind. The classical apple world is indeed a forgotten gem.

One of the earliest Greek philosophers to write about apples was Hippocrates (460–380 BC). Born on the Aegean island of Kos, he is the father of modern medicine. Everybody has heard of the Hippocratic Oath, which all trainee doctors swear by. His philosophy was very simple and down to earth. His nutritional advice was based on the presence of four humours or saps within the body, which Galen later identified as four temperaments: warm, cold, moist or dry.

Hippocrates advised eating apples at the end of meals to aid digestion and the use of cider vinegar and honey to aid the blood and circulatory system. So even if all your cider turns to cider vinegar you can put it to good use so long as you team up with a beekeeper and sell the cider vinegar in small, well-labelled, fancy bottles. You might make more money that way. 'Let food be thy medicine and medicine be thy food.'

There was a drink, mainly for the peasants, called oxycrat which sounds like an oxymoron or a low-grade government official working in the tax department in Athens. A bit sour and winey. But oxycrat was a mixture of cider vinegar, water and honey. Hippocrates prescribed it to treat wounds, sores and respiratory diseases. In Rome it was known as posca, which sounds a bit like an opera waiting to be written. The cider vinegar was useful for stopping infections in hot climates. Gladiators wounded in combat were treated with a sponge soaked in posca.

As for health and illness, Hippocrates said, 'It is far more important to know what sort of person has a disease than to know what sort of disease a person has.' Which, when applied to the cider, reads like this: 'It is far more important to know what sort of person is drinking the cider than what sort of cider the person is drinking.'

Cider was seen as a tonic, a healing force, which when applied in moderate quantities wakes up the body's own healing mechanisms. Hippocrates again: 'Natural forces within us are the true healers of disease.'

Hippocrates was credited by the disciples of Pythagoras for bringing together under one roof the disciplines of philosophy and medicine. He also managed to separate medicine from religion, believing and arguing that disease was not a punishment inflicted by the gods but, rather, the product of environmental factors, diet and living habits.

Mental illness in the ancient world was treated seriously and attempts to help certain conditions were common. If you ever lose your mind in ancient Greece or you are at a loss in your own mind, Hippocrates's aphorisms go a long way towards a common-sense approach to a cure. 'Declare the past, diagnose the present, foretell the future.'

It has to be said that Somerset psychotherapy often involves a trip to the cider house and long chats on an old sofa in a darkened barn. If Freud had lived in Somerset instead of Vienna the whole history of psychotherapy might have been very different.

Apple medicine

On a more down-to-earth level, Hippocrates advocated walking as the best medicine of all. And if, God forbid, you ever return in a bad mood, he simply advocates going for another walk. Walking in an orchard at blossom time or at harvest time is a sure remedy for all ills, as is drinking a glass of top-notch cider or perry. For you are indeed drinking the orchard, the fruits of the earth. Maybe Diogenes lived in a large cider barrel.

Then there was Democritus, another philosopher, born in Thrace the same year as Hippocrates (460 BC), a vintage year for philosophers. Democritus was the 'father of modern science' and his theory of atoms and particles has stood the test of time. Democritus held that everything was composed of 'atoms', which are physically indivisible, and that between atoms there lies empty space; atoms are indestructible, and have always been and always will be in motion; that there is an infinite number of atoms and all kinds of atoms, which differ in shape and size.

In the cider universe the same could be said for cider farms. Always it is the space *between* cider farms which has to be navigated. And when you arrive at a cider farm there should always be a blend of indestructibility and timelessness. Even cider apples are a world unto themselves. Spacing between trees is also important. An apple, Democritus reasoned, can be divided with a knife because it has air in it. More likely water as well. If the apple contained no void, it would be infinitely hard and therefore physically indivisible. Atomic cider. Divisible and indivisible.

Democritus wrote many books including: *On the Nature of Man*, *On Flesh*, *On Mind*, *On the Senses*, *On Flavours*, *On Colours*, *Causes concerned with Seeds and Plants and Fruits* and *Causes concerned with Animals*. He was very jovial and considered cheerfulness the goal of life. Democritus is often portrayed as the laughing philosopher, as if the universe is one long joke. Which it is . . .

Apples do not always get good press. Diphilus, the third-century BC physician of Siphnos in the Cyclades, simply says that 'those apples which are green and which are not yet ripe, are full of bad juice, and are bad for the stomach; but are apt to rise to the surface, and also to engender bile; and they give rise to diseases, and produce sensations of shuddering.

'Sweet apples are those with most juice because they have no great inflammatory qualities. But sharp apples have a more disagreeable and mischievous juice and are more astringent. And those that have less sweetness are still pleasant to the palate and, on account of their having some strengthening qualities, are better for the stomach. Those that ripen in the summer have inferior juice but those that ripen in the autumn have the best juice.' That is always good news for the cider maker.

A word of warning: in Greek the word 'melon' is always taken to mean apples, but 'melon' covers a number of other fruits including pomegranates. 'Armenian apples' were apricots, still grown on the northern side of Mount Ararat. 'Cydonian apples' were quinces. Cydonia was an ancient city state near Chania in Crete. Persian apples were peaches.

Just to be even more technical, the apple tribe is known in the trade as *Maleae* and belongs to the rose family *Rosaceae*. So a rosy apple is not far from the truth. Indeed, it is spot on. Cider with Rosie.

The costermonger's cry 'any old apples?' is perhaps appropriate, but it does not just apply to apples in antiquity. What is frightening is that today many apples on display in the supermarkets often come from the other end of the world and are stored for up to a year. It is not uncommon for the supermarkets to sell the tail end of one year's supply of apples from the southern hemisphere rather than stocking the freshly picked early season apples from this country. That all comes down to temperature and humidity-controlled storage, as well as strange concoctions to delay decay. I'm not sure Hippocrates or Democritus would have approved.

As to basic science and natural history Theophrastus (*c*.371–287 BC) was aware that apple seeds would not produce the same apple when planted and wrote about plant diversity in his ground-breaking book *Enquiry into Plants*. Close observation was required. There was a belief among Greek philosophers that spontaneous generation was the modus operandi of nature.

Plutarch (AD 46–120), a passionate observer of apples,

says: 'no other fruit unites the fine qualities of all fruits as does the apple. For one thing, its skin is so clean when you touch it that instead of staining the hands it perfumes them. Its taste is sweet and it is extremely delightful both to smell and look at. Thus by charming all our senses, it deserves the praise which it receives.'

Galen (AD 129–161), another Greek philosopher (Greek philosophers were ten a penny even in the Roman Empire), was also a physician and surgeon living at Pergama in modern-day Turkey not far from Smyrna, famous for its figs. For all foodstuffs, Galen said, geographical locality and season was important together with the place in which the food was consumed. Apply that to cider and you have a recipe for success. Local artisan cider, not that stuff made from imported apple concentrate. Cider always tastes best at the cider farm or in the orchard. According to Galen apples may be prepared for their sweetness and their usefulness, as well as for their suitability for storing which outranks most other fruit and vegetables. Storage measured in weeks and months, rather than days, which is a crucial point in their favour when travelling. Plus their sweetness in the old days when sugar was highly prized and only available as honey. Sugar cane was unknown in the ancient Mediterranean. It came from New Guinea via India. Cultivation began in the ninth century AD in places like Sicily under the Arabs, Andalusia in southern Spain, the Levant and Cyprus.

The sweetness of apples was very appealing, but Galen was also concerned with astringency and purgatory effects. Some apples have a cold and earthy juice which can upset the four humours, the chemical systems regulating human behaviour. Some are sweet, some are sharp, some bitter. Although outdated, this approach of the humours or saps has survived in a strange way with cider apples which are now defined as *sweet*, *sharp*, *bittersweet* and *bittersharp*. Very wisely, Galen advocates that apples should be left to ripen properly, stored carefully then cooked to reach perfection. Baked in spelt flour was a favourite.

Sappho (630–*c*.570 BC), of course, understood apples only too well when she invited the goddess of love to visit her sacred apple grove:

> At the end of the bough – its uttermost end,
> Missed by the harvesters, ripens the apple,
> Nay, not overlooked, but far out of reach,
> So with all best things.

This is from a wedding song and the ripe apple is the bride playing hard to get. Not a peach but an apple out of reach. At long last she has been plucked but not for want of striving to avoid this. A long ladder must suffice. Sappho knew a thing or two.

As to the origins of apples, Athenaeus (AD 170–223), author of the fifteen-volume *Deipnosophistae* (which translates as 'dinner-table philosophers'), asserts that it must be Dionysus, the god of the grape harvest, wine making and fertility, who discovered apples. This is attested to by Theocritus of Syracuse, who says something like: 'Storing the apples of Dionysus in the folds at my bosom, and wearing on my head white poplar, sacred bough of Heracles.'

In the *Dionysiad*, Neoptolemus the Parian records on his own authority that apples as well as all other fruits were discovered by Dionysus. So apples were indeed the fruit of the gods and those who ate them were in turn blessed by the same gods. Perhaps the breeders of apples were also acting as gods . . . although if that can be said about apples and pears, what about the cider makers?

Apples were always from the earliest times full of mystery and mystique. The spirit of the ancients lives on today if you do but look for it.

Eris and Paris

Coming as they did from Persia, the Caucasus, Armenia and Kazakhstan, apples made a very strong impression on early Greek minds. Greek myths are a rich source of apple stories which often pose interesting moral dilemmas. Seen as important catalysts, apples are waymarkers in the subterranean plots of myth and legend, where gods and goddesses, beautiful maidens and young, athletic men are caught up in a web of intrigue and decision making. Fierce animals often have to be vanquished, wrestled to the ground or outwitted. As mere mortals we can only follow in their footsteps and learn about their code of ethics second-hand and wonder at their prowess or foolishness. There is nearly always bloodshed or bloodshed averted, loss of virginity and marriage, a code of conduct, morality tales, connivance, lust and deception, where the deeper, more unsavoury characteristics of human behaviour rise to the surface and become all too visible. Universal themes, ideas and ideals that have matured in a barrel over two or three thousand years and mysteriously ended up in our own psyche.

In Greek mythology, both gods and mortals coveted the golden apples that mother goddess Gaia gave as a wedding present to Hera and Zeus. Mystical and yet forbidden. These golden apples were from the tree of knowledge in the Garden of the Hesperides, the orchard in the west tended by 'Daughters of the Evening', or 'Nymphs of the West' as they were then known.

Greeks were keen on golden apples because they not only represented health, but guaranteed eternal youth and immortality. Heracles, the illegitimate son of Zeus and the tall, beautiful Alcmene, had a go at scrumping the apples, and so did Eris, the goddess of chaos and disorder, and Hippomenes, who was rather keen on the swift huntress Atalanta.

These stories are well known. Zeus holds a banquet in celebration of the marriage of Peleus and Thetis. Peleus was from Thessaly and father of Achilles (of heel and tendon). Attending the feast were three rather special women: Hera, Athena and Aphrodite . . .

Eris then stirs things up, raises expectations and introduces a certain frisson to proceedings. Hera is the goddess of women, a matriarch, championing marriage, family and childbirth. She is sister and wife to Zeus and owns the orchard in the west. Rumour has it that Eris is her daughter so there is mother–daughter rivalry there, too. Athena is the goddess of wisdom and warfare, along with weaving and pottery. She would be the cider maker. Then there is

Aphrodite, the goddess of love, beauty and passion. Men fall at her feet.

Out of nowhere, Eris suddenly pitches up at the banquet (uninvited) with a golden apple that she has stolen from the Garden of the Hesperides. Inscribed upon the apple are the words 'For the most beautiful', which she throws at the feet of the three goddesses. As always, it is the apple asking the question.

Eris waltzes off, knowing full well that she might as well have thrown a hand grenade. Chaos ensues and the women ask Zeus to choose. But even Zeus can't fix it; he knows it will end very badly for him, so he asks Paris, a well-bred young lad, to choose the most beautiful of the three.

The story now goes that the three women were led by Hermes to a spring on Mount Ida where they bathe naked. Paris just happens to be passing and looks on at the beautiful scene so that he can make his choice. He is entranced by what he sees. All three, seeing him hovering, offer him gifts that amount to bribes, to sway his mind and aid his discerning judgement.

Hera offers to make him King of Europe and Asia Minor. Athena is slightly more subtle and offers him wisdom and skill in battle. But Aphrodite, knowing what men really want, offers to give to him the love of the world's most beautiful woman: Helen of Sparta, married to King Menelaus. Paris chooses Aphrodite, a decision that caused the Trojan War and ultimately the destruction of both Paris and his city, Troy. It was a ghastly business. So apples and scrumping can affect the course of history. They are deeply embedded in cider culture.

Orchard memory

Enter Homer, without whom we would have very few Greek stories. Homer loves a good tale: lots of young men with swords and shields and boats roaming around the Aegean, Middle East and Black Sea, chasing wives, golden fleeces, minotaurs, boars and mistresses. A few battles, endless trials and tribulations and then the homecoming. Slightly on the run, Odysseus comes across King Alcinous living on an island, possibly Sicily or even Corfu. In Greek Alcinous means 'mighty mind'. He has a large palace and a fine 4-acre orchard surrounded by a hedge. The estate agent's blurb runs like this:

> Tall, heavily laden trees grow there, pomegranate, pear and apple, sweet figs and dense olives. Winter or summer the fruit never rots or fails. It lasts all year (which is a miracle) and the West Wind's breath quickens some to life, and ripens others: pear follows on from pear, apple on apple, cluster on cluster of grapes, and fig on fig. A continuous feast. There is the vineyard with a warm patch of level ground for drying grapes, while the labourers gather, others tread, and as the foremost rows of unripe grapes shed their blossom, others become tinged with purple.

Paradise indeed.

When Odysseus eventually returns home to the island of Ithaca, he visits his father Laertes in his orchard. He sends his servants into the house so as to be alone with his father. Laertes has grown old and rustic. He dresses like a peasant and has aged prematurely out of grief for the loss of his son and wife. He doesn't recognize Odysseus, but Odysseus doesn't say who he is and pretends to be someone who once befriended Odysseus. Laertes begins to cry at the memory of his son. Odysseus then throws his arms around Laertes and kisses him. He proves his identity by pointing to the scar on his thigh caused by a boar's tusk and with memories of planting fruit trees with Laertes when he was a boy. Thirteen pear trees, ten apple trees and forty fig trees in the orchard. It is all about orchard memory, planting ideas and fruit reunited. The orchard is a symbol of past recognition, reconciliation and a place to contemplate life after the turbulence of war and travelling. The same applies today . . .

Another Greek apple story concerns a young man called Hippomenes. A brave lad but, not wanting to be killed, he asks Aphrodite (who won the beauty contest sparked by the golden apple) for advice on how to win.

The story goes that Aphrodite (who was very sensual herself) disapproved of Atalanta who spurned love in favour of hunting and winning races. Perhaps Aphrodite could not bear to see so many young, strong, athletic men put to death for not running fast enough, for she advised Hippomenes to take with him three golden apples. How she had come by them is not revealed. Maybe she paid a visit to the Garden of Hesperides on a moonlit night. Ovid reckoned the orchard was in Tamassus, 20 kilometres south of Nicosia. Cyprus again.

As for the race, Atalanta was so confident that she gave Hippomenes a head start and, each time she was about to overtake him, he dropped a golden apple, so distracting her that she slowed down in order to pick up the apples. In the end Hippomenes won and married the girl. They had a son. But like all Greek stories there is sting in the tail. It does not end happily for they are so besotted with each other that they have sex in a temple and defile the sacred space. The price they pay is to be turned into lions, some say by Zeus or Artemis, others by Aphrodite herself who was annoyed with Hippomenes for not showing enough gratitude for her help. Even gods get jealous.

In the cider counties apple picking was on many farms and still is, a major sorce of income for young women. Hard work on your knees all day. Apple pickers are notorious for their earthy sense of humour.

Apple throwing

Apple throwing is not an Olympic sport, but it was admired in ancient Greece. Apples and philosophy were often flavour of the month in the autumn after the long, dry heat of summer. Breeding apples, grafting and propagating them was a kind of sport, a way to show off to your neighbours. Alexander the Great (356–323 BC) was in it up to the hilt. He sent back dwarf apple rootstocks from Persian apple orchards to the Lyceum in Athens. His philosophy teacher Aristotle had encouraged him to go east in the first place and attack Persia. Maybe Alexander wanted to find out where apples came from. He got very close to the source with his forays into Scythia, Bactria and Sogdia, north of Afghanistan. It was this great interest in the natural world which propelled Alexander forward as much as the desire for conquest.

Is it the apple you desire or the thought of an apple or the associated memory of sensation when eating an apple? Taste, touch, smell, texture, delight, sound, even the crisp crunch. Apples in the end rot; like us they are transitory. But apples live on in cider for a year or two and within cider's distilled spirit can live for decades, even a century or more bottled up. Impermanence and immortality in one fruit. No wonder the Greeks loved apples.

As for Alexander's dwarf apple rootstock (which he sent back to Aristotle[1]) it was later named *Malus pumila*, the small apple. Maybe it lasted till the spring, like the May Queen? Very useful on campaign or a long voyage. So Alexander was a fine but ruthless example of a military plant hunter.

There are many different types of dwarf rootstocks, among them: French Paradise, *Malus paradisiaca*, and the more vigorous English Paradise. One variation of this rootstock was recorded in Paris 1398 at the Hôtel St Pol, and by the sixteenth century Paradise apple was recommended as a dwarf stock. It is known today as M9, not a motorway in Scotland but a recognized variety from East Malling in Kent. Just to confuse matters, the M27 is for very small

1 It is well known that Aristotle was keen on apples, so much so that *The Book of the Apple* was attributed to him and discusses his own ideas about immortality. As he lies dying, he is periodically revived and energized by being offered an apple to smell. The reference is in Arabic: *Risālat al-Tuffāha* or *Tractatus de pomo et morte incliti principis philosophorum Aristotelis*, allegedly a medieval neoplatonic Arabic work of unknown authorship.

almost bonzai apple trees and M26 for semi-dwarfing. The full-blown M25 is a real cider rootstock. Heavy cropping standards. In fact, there are now over fourteen different types of Paradise apple. Alexander would have been proud.

The Greeks in Athens found dwarf trees very useful in their gardens where much of their philosophy was debated. The garden was in effect library and classroom. Sitting under the shade of an apple tree must have been a good way to spend a morning. As with all students, apples had a raunchier edge to them after hours: there was more to the expression 'throwing a party' than meets the eye.

Apples were also an important part of Athenian wedding feasts. Solon (630–560 BC) decreed that during marriage ceremonies and before entering the bridal chamber the bridal couple should share an apple before going to bed: 'In order that the first meeting be not disagreeable nor unpleasant, bride and bridegroom may be shut into a chamber and there eat an apple together.'

Maybe they had a whole bowl of apples to get through. One bite of the apple – a lifetime of marriage. Some say it was a quince or Cydonian apple but eating a raw quince is not easy.

Also in ancient Greece tossing an apple to a woman was akin to a proposal of marriage, catching, an acceptance. In *Clouds*, a Greek comedy of ideas, Aristophanes (460–380 BC) states: 'throwing apples was a ploy at seduction, since the apple is sacred to Aphrodite.' Another version was a warning to boyhood innocence and shows that girls were also throwing apples. 'And don't run after dancing girls, so that as a consequence, you don't get hit by an apple thrown by a prostitute and wreck your good name completely.' So if you were singled out and pelted with apples in public by lively women your reputation and marriage chances were ruined. This puts the apple on a different footing and evens up the score. Here the apple, and therefore the power, is very much in the woman's hand.

Alexander's apple

Even today apples are often thrown around cider farms by young men and women to attract someone's attention. Small cider apples (both inexpensive and biodegradable) fit nicely in the palm of the hand and are good for target practice. And even if you hit someone you rarely 'do them harm'. Some people just throw apple cores, having eaten the fruit. Throwing apples at weddings is somehow preserved in the act of throwing confetti which resembles apple blossom when the happy couple come out of the church. So the apples are in kit form. In Serbia there is a custom whereby an apple is hung at the top of a tree outside the bride's house and only when the groom has shot the apple down can he enter her house.

Throwing apples was quite a pastime in Greece and indirectly led to the death of a senior Greek army officer in 328 BC. Alexander the Great threw apples as well as tantrums, notably when feasting with Cleitus the Black, who had saved his life at the Battle of Granicus, near Troy. Alexander was very grateful to him for many years. Cleitus's sister Lanike had also been Alexander's wet nurse, so

Cleitus was almost family. Things progressed well and Alexander fought various tough battles when on campaign around Central Asia. Six years later Cleitus was offered control of Bactria as satrap, or provincial governor.

But just as this was about to happen, Alexander ordered a feast in Samarkand. One hopes cider wasn't involved, but fighting broke out between the guests. Alexander then changed his mind and gave Cleitus command of 16,000 defeated Greek mercenaries who had fought against him for the Persian king. He then asked Cleitus to fight the warlike nomads who lived on the Central Asian steppe, an interesting task if ever there was one. Good Turcoman carpets were to be had. But Cleitus took insult at this demotion and command of what he regarded as second-rate troops. An argument ensued. Cleitus may have taunted him about his sister being his wet nurse. Alexander saw

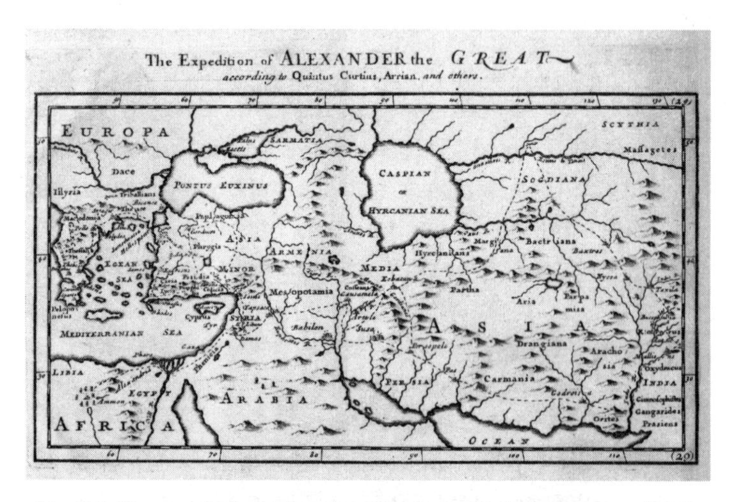

Fig. 2.1 'Route of Alexander the Great's apple hunting expedition to the Middle East and Central Asia', *c.*334–323 BC

red and threw an apple at Cleitus, which hit him on the head, and then called for dagger or a spear. Alexander was restrained but the gloves were off. More insults were traded, the wine was talking and, later on, Alexander grabbed a javelin and threw it into Cleitus's heart. When his friend died, Alexander was grief-stricken, recalling that Cleitus had once saved his life and that he was also his nurse's brother. He was indebted to their family. Five years later Alexander was himself dead. Apples have a way of providing hidden messages. Core truth.

It was Plutarch who first recorded this particular apple-throwing incident, but there were other apple episodes. Anaxarchus, on being pelted with apples by Alexander at a dinner party, stood up to retaliate and threw one or two quips back in his direction, saying, 'A god shall take a hit from mortal hand'. Anaxarchus laughed at Alexander for making himself a god. Everyone had to bow down. Not a good sign.

Historians of Alexander also mention an elite Persian regiment, 10,000 troops, called *Immortals*, or *Apple Bearers*. Their spears had a small metal counterweight to balance the heavy point which resembled an apple. Herodotus (484–425 BC) refers to the Immortals in his description of Thermopylae, fought 150 years earlier in 480 BC. Xerxes, the Persian King of Kings, had an elite bodyguard, a thousand men with spears, known as *arstibara*, nicknamed 'Apple Bearers'. Apples were everywhere. Even on the Persian front line.

Anaxarchus from Abdera studied philosophy with Democritus and was a friend of Alexander. Plutarch reports that he once told Alexander that there were an infinite number of worlds. Alexander became dejected because he had not yet conquered even one.

But to return to Mesopotamia, Alexander, mighty warrior and tactician, eventually came unstuck. Location: Nebuchadnezzar's palace, Babylon. A drinking binge. June 323 BC, about six years after the apple-throwing incident with Cleitus the Black. Alexander was pipped to the post. Towards the end of the palatial drinking session, which had gone on all day, he collapsed and lost his voice. He lingered for a fortnight before dying aged 32. Humans who think they are gods are in the end mortal. Some say he was poisoned.

Specific gravity

But the Greeks were not just philosophers and poets. They were problem solvers and practical scientists. Take buoyancy. Apples float, which means you can move them around your farm yard by using a heavy duty hose to power wash them into narrow channels that lead to the cider press. Apple canals. During a storm, apples in an orchard will sometimes flow across the grass downhill and end up in a ditch or river. Brook apples.

As a cider maker one of the things you need to find out in the autumn is how much natural sugar is in your apple juice. Here Archimedes (287– *c.*212 BC) of Syracuse in Sicily comes to the rescue. His principles of floating bodies rule the liquid universe. Fluid mathematics. What goes down into water must displace an equal volume of liquid to its own weight. That is the Archimedes principle and this can be used to measure the density of apple juice.

The more sugar in solution the denser the juice. These days cider makers use a hydrometer, a finely calibrated instrument made of glass with a thin top and bulbous bottom that is weighted down. You pour apple juice into a tall glass tube and drop the hydrometer in carefully. Spin it a little, let it bob up and down then wait till it settles. Then take readings, which will be around 1.050. That will give you about 6.3 per cent alcohol, which is fine, pure water being 1.000. In a really good summer sugar levels can go up to about 1.070, i.e. 9 per cent abv (alcohol by volume). Reading a meniscus is not easy. You have to take the bottom of the curve and see how it sits on the juice. Meniscus sounds like yet another philosopher, but it is in fact Greek for 'crescent', as in crescent moon.

The crucial thing is that the specific gravity, known in the trade as SG, gives a clear indication of how strong your cider will be after fermentation. The more sugar, the denser the apple juice, the more alcohol at the end of the day. A wondrous metamorphosis. Nature's gift that keeps cider makers on their toes. Buoyancy matters.

And although Archimedes did not invent the hydrometer his theorems helped it on its way. He also invented the Archimedes screw, which is very useful for lifting water from the Nile to the fields of Egypt and for offloading grain in warehouses and on farms. More importantly, the Archimedes screw is used for loading cider apples into trailers and dealing with spent pomace, the dryish apple residue that is left after squeezing all the juice out of it. Pomace is best spread on the fields. Sheep and cattle love it. Pheasants, too. Even Bulmers' new factory in Ledbury uses a large Archimedes screw for moving apples.

Archimedes was also an expert in using levers for moving large weights like boulders and full cider barrels. To every action there is equal and opposite reaction. Many of his thoughts are still with us, and very useful for cider makers.

Sadly, Archimedes died at the hands of an irate Roman soldier during the Second Punic War, after a two-year siege. The soldier mistook his mathematical instruments for weapons and tried to seize them. Archimedes quite rightly resisted. The soldiers were under orders that he should not be killed. His last words were 'Do not disturb my circles'. He was in the middle of yet another geometric theorem.

Hydrometers

Then there is the hydrometer, that delicate, clever little glass instrument used to test the apple juice and cider once it has fermented. The earliest ones were perfected by a brilliant woman from Alexandria called Hypatia (*c.*AD 360–415). She was a philosopher, inventor, mathematician and astronomer, a Hellenistic neoplatonist pagan. A great thinker and a remarkable woman. We know about her because of a letter written to her by one of her disciples,

Synesius, a Greek bishop in Cyrene. In letter No. 15 Synesius requested that she send him a hydrometer to help him with his experiments. She also built astrolabes.

Hypatia was a very unusual woman. Often consulted on politics and philosophy, she was held in very high regard but became embroiled in a political feud between Orestes, the Christian prefect of Alexandria, and Cyril of Alexandria the vicious bishop. It concerned the role of church and state and whose authority governed the land. Sound familiar? Hypatia sided with Orestes and suffered a ghastly death. She was seized by a mob when on her way home in a carriage then dragged to a pagan temple, stripped and stabbed with shards of pottery. Her mangled body was then burned. So whenever you squint through a hydrometer looking at the meniscus, think of Hypatia and Archimedes. Tragic ends to brilliant careers.

Apple dining clubs

Imagine you are in Italy in the first century. Pliny (AD 23–79) notes in his *Natural History* that there are thirty-nine varieties of pear and twenty-three of apple. Pliny was an author, natural philosopher, naval and army commander. He died, some say of noxious fumes, when Vesuvius erupted, bravely trying to rescue his friend Rectina and the library in her villa. Pliny was in charge of the Roman navy at the time and sailed back towards Pompeii from Misenum on the other side of the Bay of Naples just to see what was going on. Others say he died of a heart attack and was nowhere near Vesuvius. While the circumstances around his death are uncertain, his observations on the natural world are very pertinent.

Apples listed by Pliny include the Greek *Epirotic* apple, the *Syrian Red*, *Matian* apples, *Scaudian* apples and a small *Petisian*. Then *Honey* apples, *Flour* apples, the *Little Greek*, the *Armeia* after the town of Amelia in Umbria and late keepers like the ruddy *Apian*, which smelt of quinces.

The *Pomme d'Api* was brought from the Peloponnese to Rome by Claudius Appius in the fourth century BC. Pliny also mentions 'wild apples with fine flavour, peculiar pungency, or such acidity that they could blunt a sword blade', and 'Fruits That Have Been Recently Introduced', including *Cestian*, *Mallian*, *Amerinian*, *Græculan*, *Gemella*, *Melapium*, *Musteum*, *Melimelum*, *Orbiculatum*, *Orthomastium*, *Spadonium*, *Melofolium*, *Pannuceum*, *Pulmoneum*, and *Farinacean*.

We are therefore very grateful to Greek and Roman historians and poets. When the Renaissance occurred 1,500 years later, it was their work that English scholars turned to, and through their eyes saw the natural world around them. From Greek to Latin to Arabic then back to Latin and into English and French – a remarkable survival of knowledge.

Pomariums are mentioned by Pliny and the custom of auctioning off fruit still on the apple trees. About storing apples Varro (116–27 BC), one of Rome's greatest scholars and author of *De re rustica*, says this:

> The pome fruits, like preserving sparrow apples, quinces and varieties of apples known as Scantian, and 'little rounds' (*orbiculata*) and those called winesap (*mustea*), now called honey apples (*melimela*), can be kept safely in a cold, dry place when laid on straw, and those who build fruit houses take care to have the windows give upon the north wind and that it may blow through them: but they should not be left without shutters for fear that the fruits should lose their moisture and become shrivelled by the effect of the continuous wind.
>
> So pleasing were these *pomariums* that the Romans liked to dine in them. Apple dining clubs were the height of society. The vaults, walls and pavements of these fruiteries are usually laid in stucco to keep them cool: thus rendering them such pleasant resorts that some men even spread there, their dining couches: as well they may, for if the pursuit of luxury impels some of us to turn our dining rooms into picture galleries in order to regale even our eyes with works of art [while we eat], should we not find still greater gratification in contemplating the works of nature displayed in a savoury array of beautiful fruits, especially if this was not procured, as has been done, by setting up in your fruitery on the occasion of a party a supply of fruit purchased for the purpose in town?

Juvenal (second century AD) issues an invitation to his guests and regales them with an offer of fruits in Satire XI, *Extravagance and Simplicity of Living*:

> There will be grapes too, kept half the year, as fresh as when they hung upon the tree; pears from Signia and Syria, and in baskets fresh-smelling apples that rival those of Picenum, of which you need not be afraid, seeing that winter's cold has dried up their autumnal juice, and removed the perils of unripeness.

Apicius (first century AD), a noted gourmand, gave his name to a book of Roman recipes, which has survived the ravages of time. Roman pork with Matian apples. Chopped apples with yogurt. Patina of pears: 'Core and boil the pears, pound them with pepper, cumin, honey, passum, liquamen, and a little oil. Add eggs to make a patina, sprinkle with pepper and serve.'

To many Greeks, including followers of Pythagoras, the apple was a symbol of the occult. If you take a knife to it vertically it splits in half and then quarters, which show the pips and core in a dissected state. If, however, you take the knife and cut the apple horizontally, the apple depicts a perfect five-point star, the pentagram, the key to knowledge of good and evil.

Pomona

Romans were keen on apples and orchards; they knew how to prune, graft and propagate apple trees. It was the height of luxury and a mark of status to have a well-maintained orchard if you were a retired general, senator or just a wealthy farmer.

Pomona, the 'orchard pruner', was a Roman wood nymph associated with fruit and fruit trees, gardens and orchards. Her badge of office was a small, curved pruning hook. Over the centuries she became associated with apples to such an extent that illustrated books on apples are called *pomonas*, e.g. *The Herefordshire Pomona*, *The Somerset Pomona* and *The Cornish Pomona*.

Pomona was the hard working goddess of fruit and orchards. She lived in Roman orchards and helped to maintain them. She was one of the friendly *genius loci* and often depicted as a young, graceful maiden. These *genius locii* were not necessarily immortal, but lived many years before they died. The Romans then brought them to Britain where they fitted in very neatly with the Celtic beliefs of local spirits. In Greece, these nymphs included the Hesperides who presided over evening sunsets as well as the orchard of Hera. But Pomona does not have an exact Greek counterpart. Roman through and through, she silently protected fruit trees and cared for their cultivation.

Pomona used her curved pruning knife to cut back luxuriant growth and lopped branches that spread out, splitting the bark and inserting a graft, providing sap from a different stock for the nursling. She would not allow them to be parched, watering, in trickling streams, the twining tendrils of thirsty root. This was her passion. She had no longing for desire. But fearing boorish aggression, she shut herself away in an orchard and shunned men.

In the myth by Ovid she scorned the love of the woodland god Silvanus and Picus, the King of Latium. Silvanus presided over plantations and delighted in trees growing wild. He also protected cattle and warded off wolves. Worshipped by shepherds, he was a keen musician associated with Pan. But he was not good enough for Pomona. She was very particular.

Pomona was tricked into marriage by Vertumnus, the god of seasons, change and plant growth, gardens and fruit trees. An ideal companion for Pomona you might think. Vertumnus, an Etruscan god, could change his form at will. A bit of a trickster.

By his many disguises, he gained admittance and enjoyed gazing at her beauty. Once he covered his head with a coloured scarf and, leaning on a staff, with a grey wig, pretended to be an old woman. He entered the well-tended garden, admired the fruit, and said, 'You are so much more lovely', and gave her a few kisses, as no old woman would have done. He sat on the flattened grass, looking at the branches bending, weighed down with autumn fruit.

Thus Vertumnus gained Pomona's confidence and engaged her in conversation: 'Put aside, I beg you, reluctant pride, and yield to your lover. Then the frost will not sear your apples in the bud, nor the storm winds scatter them in flower.'

Slowly, Pomona succumbed to Vertumnus's advances and, suddenly he became a youth again. He threw off the dress of the old woman and appeared to Pomona as a radiant sun when it comes out from behind a cloud. Vertumnus, being an opportunist, was ready to take her, but no force was needed. The nymph was dazzled by the new appearance of this warm god and felt a mutual passion, a stirring from within, and yielded to his advances of her own accord. Or so we are told . . .

So the god of seasons and goddess of orchards were finally grafted to each other, joined at the hip, which makes perfect sense. The festival of Vertumnus and Pomona was celebrated on 13 August, just about when the earliest apples first appear. Not bad timing. After all, orchards are at the mercy of the seasons. It is the seasons that give the drive and inclination to the fruit. The match was perfect, and they lived happily ever after.

In Emerita Augusta in Spain there is a mosaic colourfully depicting Pomona with fruit wrapped around her head in the form of a diadem and bearing the word Autumnus. This can be found at the fourth-century AD Villa Las Tiendas, near modern-day Merida, a town established for retired soldiers who often brought fruit trees from remote regions where they had been serving. They also took grafts with them to the remote places like Britain to remind them of home, and that is how many of the larger, sweeter apples ended up on our shores. Often where vines struggled in the colder, wetter climate, apple trees blossomed. Pomona came to Britain and stayed.

Recent excavations on Hadrian's Wall at Vindolanda have unearthed a tablet which mentions apples in what looks like a shopping list:

Tablet 302

bruised beans, two *modii*, chickens, twenty, a hundred apples, if you can find nice ones, a hundred or two hundred eggs, if they are for sale there at a fair price. . . . 8 *sextarii* of fish-sauce . . . a *modius* of olives.

And, just to clinch it, there is a mosaic in Cirencester which was discovered in 1849 by workmen digging a drain. It shows Pomona with her pruning hook rising up out of the underworld. Orchards were here to stay. Pomona now had her work cut out.

Fig. 2.2 'Autumnus, Pomona's sister', Roman mosaic in a bath house, Villa Las Tiendas, Spain, 4th century AD

3

APPLES OF THE IMAGINATION

Apples blossom in our imagination. We first come across them in children's stories of poisoned apples, witches, snow queens and kisses from princes. Apples are used by gods and goddesses to hold on to their youthfulness and by strong, beautiful women to lure unwary men away on long, dangerous voyages to distant lands. Apples are a symbol of fertility and hope, even initiation.

What is interesting about Celtic, Irish and Norse apple legends is that they seem to occur independently of Roman/Greek traditions. They may well be the result of earlier more northerly migrations into Europe. There are several academic theories about the various routes taken, but sadly nomads leave few clues. These migrations began in Central Asia about eight thousand years ago, just as the steppe was opening up after the last Ice Age. Stories travel just as well as apples and are often interlinked, which is why they are so important.

Apple stories are particularly widespread. There was a

more southerly route for apples through the classical world and then a more northerly one into Scandinavia via the Baltic. These colourful apple myths and sagas constitute a rich seam within our own culture and are not just taken from the Greeks and Romans. Many of these beautiful myths have come down to us from Sweden, Norway, Iceland, even Germany. Others are deeply embedded in the Celtic lands of Ireland, Wales and Scotland, always with a slightly different twist, but often heroic.

Danger often lurks beneath that beguiling beauty, none more so than in the story of William Tell. Bows, arrows, quivers, and young boys. The need to be deadly accurate. This apple story first emerges from a Danish source, Saxo Grammaticus (1160–1220), in the twelfth century, then, in the thirteenth century, in the saga of Þiðrekr of Bern, also known as the *Vilkina saga*, an Old Norse chivalric tale. These stories go round and round. The text is probably by a Norwegian scholar from the 1200s who translated a lost Low German prose narrative, although it could also be from an Icelandic scholar from the 1300s. These sagas were often told in winter round the fire. They still occupy the minds of modern-day scholars today.

Apples always tell a story, which is why they are so important. There is a hidden truth to these images which still resonates within us. Myths embody the human experience on a divine level and sagas are more often oral accounts, prose stories and structured narratives that are told and retold thousands of times. Traditional sagas usually concern super-human beings or heroes embroiled in a natural or semi-mythical world populated by deities or demigods. These stories can also inform the background for rites of passage and strange natural phenomena.

What is intriguing is that apple stories are still being retold and reinvented. Take for instance the Anglo-Saxon tale *Beowulf*, as adapted in 1968 by the English poet Robert Nye especially for children. There are alas no apples in the original *Beowulf* manuscript but Nye adds an encounter with an apple tree on the way to the Great Hall when Beowulf meets Hrothgar, ruler of Denmark, and Unferth, his thane. When Beowulf offers them each an apple, neither Unferth nor Hrothgar will eat the fruit owing to superstition. They are afraid, claiming that the fruit grew from the green, rotting teeth of an old witch: 'An old witch spat her teeth out there. They were bad teeth – green and red and rotting. They grew into apple trees. Nobody in his right mind would eat fruit like that.' It must have been a very fine, high-tannin crab apple.

But Beowulf is not put off by their recalcitrance and eats the apple all the same. Unferth suggests that only a wicked person could eat bewitched apples and survive, and Beowulf laughs. He tells Unferth that 'bad does not always come from bad'. Beowulf opens his mouth and shows him his rotting teeth. This puts apples in a different light. Although you do need good front teeth to bite into apples . . .

A more modern twist may be seen in the death of Alan Turing, the celebrated mathematician, computer scientist and codebreaker who is said to have killed himself by eating an apple laced with cyanide. The apple was found beside his body half eaten. The contemporary American novelist David Leavitt has speculated that Turing was re-enacting a particular scene from the Walt Disney film *Snow White and the Seven Dwarfs* (1937), Turing's favourite fairy tale. The scene involves the Wicked Queen who immerses her apple in the poisonous brew. Others noted that Turing was conducting experiments with cyanide to electroplate spoons with gold in his side room. Turing often ate an apple before bedtime. An apple enigma.

This is perhaps why some people believe that the Apple computer logo, an apple with a single bite out of it, is a tribute to Turing. A modern myth, but a powerful one which shows how durable and elastic apple myths really can be. Apple denies any knowledge. Just a wonderful coincidence.

Then again, in contrast to the poisonous, apples there are countless golden apple stories. One intriguing link is to be found in the Nart sagas from the Caucasus, a valuable storytelling stepping stone between Central Asia and Europe. The Narts were descended from proto Indo-Iranian people, who had come out of Central Asia. They had links with nomadic Kazakhs and the Scythians who traded so successfully with the ancient Greeks. The Scythians were very fine artists in gold and wonderful examples exist of golden apple trees and horses. Some of these apple stories must have gone north round the Black Sea deep into modern-day Russia and Ukraine and later appeared in the Baltic regions.

In the Nart sagas there is a magical golden apple tree which produces blossom in the morning and ripe fruit by evening. One side of the fruit is red, the other white. Then fertility comes into it, as it always does. If a barren woman eats the white side, she will have a daughter with white, silken hair. If a barren woman tastes the red side, she will have a great son, also with silken hair. But then the Narts found that, mysteriously, the apples were being stolen, so the tree was guarded day and night by two brothers. One day they discovered three doves taking the apples so one of the brothers fired an arrow and wounded one of the birds. The doves flew off, taking a golden apple with them, but the blood from the wounded dove fell on the ground. The brother who fired the arrow dabbed his handkerchief on the blood and they then followed the trail round the Black Sea.

When they came to the Sea of Azof (Azoff) they cut a path through the water with a sword and eventually came upon a white house in a misty ravine occupied by seven brothers and three sisters, all descended from the goddess of water. There was a fine array of food on offer, including a stolen golden apple. They knew they were on the right path. Only two sisters greeted them; the third was lying down and dying slowly of the wounds she had received as the dove. One of the brothers took from his pocket the bloodstained handkerchief, moistened it and dabbed her wound, whereupon she miraculously recovered.

The sisters said that they would put on the skins of doves and fly south to the tree of golden apples to find husbands. And so it was that the wounded 'dove' married the man who had wounded her and then brought her back to life. This intriguing apple tale is taken from John Colarusso's book *Nart Sagas: Ancient Myths and Legends of the Circassians and Abkhazians* . . . So there you have it: a golden apple tree, theft, blood, birds, flight, beautiful young women, strong men, migration, feasting and marriage.

Norse apple sagas

All these same apple elements wove their way into the myths of Scandinavia and even Ireland. Maybe it is no coincidence that large parts of Scotland and northern England were under the Viking yoke for generations. Certainly in Norse myths apples were held in high regard in the early tenth century AD when on these shores people spoke Old West Norse alongside Old English. Many words crossed over.

Scandinavia has its own much earlier ancient apple culture. Remains of dried apples have been found in Sweden at Östergötland and have been dated to *c.*2500 BC. In Norway the remains of seafaring apples have been discovered. In the Oseberg ship, dating to about AD 820, a very ornate bucket was found containing wild crab apples, grave goods for the journey to the afterlife. Cut in half or quartered, crab apples are easy to store and become more digestible as time passes. Dried apples are good for sea voyages.

Even in Iceland, which was settled from Norway in the eighth century AD, there is a popular goddess called Idunna who is associated with apples and eternal youth. Apples must have been a rare luxury in latitudes close to the Arctic Circle, unless that is the climate was better in the past. Often the idea of an apple is as strong as the real thing.

As it happens, an Icelandic neighbour called Sigrun Appleby used to live in my village. Her Icelandic family name is Hólmsteinsdóttir, from Reykjavík. Her uncle Sæmi is one of the very few people to grow apples in Iceland. This has taken time and patience. He was a schoolmaster and chose Norwegian apples like Savstaholm and Haugmana and Canadian varieties like Carol and Mantet as well as English varieties Emneth Early and George Neal. The sooner the apples ripen under the midnight sun the better. Sæmi – Sæmundur Guðmundsson – grows them close to his house, facing south where he can keep an eye on them. Sæmi lives 70 miles east of Reykjavík in Hella, set in a fertile river valley which runs southwest, an ancient part of the country where many Icelandic sagas are set. Hekla is an active volcano believed to be the gateway to Hell.

Just west of Hella is a farm with twelve caves once inhabited by Irish monks. To the south is another farm called Oddi, the site of an ancient monastery where two of the most famous Icelandic sagas of the thirteenth century were believed to have been written: Snorri Sturluson's *Prose Edda* and Saemundur Sigfusson's *Poetic Edda*. So apples may well have grown there in the past. The heart of Icelandic culture: horses, poems, salmon, sagas and apples.

Ever young, the goddess Idunna is always seen with her basket of apples by her side. Hilda Ellis Davidson, who studied pagan Scandinavian mythology at Cambridge

University, found several references to apples in the sagas. In *The Prose Edda* book *Gylfaginning*, Idunna is described as Bragi's wife and keeper of an *eski*, a wooden box made of ash wood and used for personal possessions, within which she keeps apples. The apples are bitten into by the gods of Asgard when they begin to grow old and they become young again. Immortality regained with apples given by a beautiful young woman. Same story. Same theme.

But there is always trouble lurking. The giant Þjazi forcibly takes Idduna far away from Asgard. She is lured into a forest by the god Loki who tells her he has found some new apples that she might be interested in and she should bring her own apples with her to compare them. A bit of pomology and apple identification on the quiet. No wonder she cannot resist. But then Þjazi appears as an eagle, grabs Idunna and flies away with her to his realm in Jotunheim.

This abduction and the loss of the apples causes the premature ageing of the gods who become angry and threaten the giant with torture and death. But to no avail. Loki transforms himself into a falcon, flies to Jotunheim and finds Idunna, who he turns into a nut before flying back with her in his beak. When Þjazi discovers she has gone he flies towards Asgard but the gods light a fire which singes his feathers and he falls to his death. His eyes become two very bright stars. The apples and Idunna are rescued.

Hilda Davidson also notes that eleven 'golden apples' were given to woo the beautiful Gerðr by Skírnir, who was acting as messenger for the major Vanir god Freyr. In the *Skírnismál*, also from the *Edda*, Davidson observes a

Fig. 3.1 'Freia: a combination of Freyja. Goddess of love, beauty and guardian of the golden apples and the goddess Iðunn', 1910

connection between fertility and apples in the *Völsunga saga* when the major goddess Frigg (who gives us the name for Friday, Friggsday) sends King Rerir an apple after he prays to Odin for a child. Frigg's messenger, in the guise of a crow, drops the apple in his lap as he sits atop a mound. Rerir's wife eats the apple and this results in a six-year pregnancy and the caesarean birth of their son – the hero Völsung. Fertility, but at great cost and the death of the child's mother.

Germany and Holland also have their own pagan pre-Christian goddess Nehalennia, dispensing apples to those undertaking voyages across the North Sea. Temples have been found in Holland dedicated to her. On stone altars she is portrayed with apples and loaves of bread like a mother goddess and a hunting dog. Very common symbols at this time. Little is known about Nehalennia. Her name has also proved difficult to crack. One recent theory put forward by the Italian scholar Patrizia de Bernardo Stempel links the name Nehalennia with Welsh *halein*, 'salt', and *heli*, 'sea', proposing a Celtic origin. She deconstructs the name as a combination of Celtic *halen*, 'sea' and *ne-*, 'on, at'. Finally, *-ja* is a suffix forming a feminine noun. Thus, the meaning would be 'she who is at the sea', whether it be the Irish Sea, North Sea, Baltic or Denmark Strait.

Seafaring was very much a Viking trait. In AD 997 Vikings raided two very good cider areas in South West England, first Watchet in Somerset, and then they sailed round Land's End and up the River Tamar. They beached their vessels near Hole's Hole, before heading over the hill to

Tavistock, where cider definitely was made. Here they sacked the abbey, drank the cider and raided the mint at Lydford. Alas, no apples. Silver Lydford pennies have been found in Sweden, Denmark and Gotland in the Baltic, which shows the distances covered by the raiders. They may well have been used to pay Danegeld. Today cider is made in Sweden in Kivik, where there are old orchards, and also near Malmo, where there is one cider maker from Gotland. These places were well connected in the past by elegant sailing vessels like the one unearthed at Sutton Hoo in Suffolk in 1939.

So we see that the traditions of Norwegian and Swedish apples are nothing new. Documentation goes back to the thirteenth century when Cistercian monks introduced apple growing. Norwegian cider production peaked in the late nineteenth century and officially stopped in the 1920s with prohibition, as it did in the United States. Now, a hundred years later, it is getting going again. Long summer days. Perfectly good cider is now being made in Hardangerfjord and Sognefjord – the fjords provide very good microclimates for apples.

Like Canada, Sweden also makes ice cider these days. This is not to be confused with that awful sweet, sickly stuff called pear cider, made with pear concentrate. Thank you, Kopparberg. Artisanal cider is now a force to be reckoned with. Idunna would be pleased. Anything to keep the Norse gods happy.

What is curious is the similarity between Greek apple myths, Celtic apple legends and Norse apple sagas. All seem to come from the same source of Indo-European myths, Central Asia, which is where the apples originate, but via different routes, one north, one southern. That is where the Nart sagas from the Caucasus fit in, a nomadic apple myth like the apple itself that has survived and diversified for thousands of years. Apples had the capacity to migrate westwards and this has been mirrored in the slow shift from nomadic pastoralism to more settled agriculture. These important stories are the fruit of many journeys both physical and metaphysical. They are at the centre of our belief system and they underpin the main tenets of cider philosophy. First the apple, then legend myth and saga, followed by cider making and philosophy distilled.

Apple archaeology

And that reality is embodied in archaeology. Early evidence of apples is not just restricted to the Middle East. Remains of crab apples and larger cultivated apples have been found in Bronze Age sites throughout Europe, such as La Tène prehistoric lake dwellings in Switzerland and at Hallstatt in Austria.

In Devon, at Hazard Hill, near Totnes there is evidence of Neolithic people collecting crab apples for eating, cooking or even cider making. And at Hayes Farm, Clyst Honiton, *c.*3750 BC. Crab apples (*Malus sylvestris*) are often small and hard, so they need plenty of squashing if they are to be used for cider. Often cooked on a stone, they were part of our diet 5,750 years ago. Archaeology is all about food history with sacrifice and agriculture thrown in. Today crab apples are planted as key pollinators in the hedges of orchards.

To find more about apples in ancient towns I contacted the archaeologist Peter Addyman in York who kindly gave me a list of places in York where apple remains have been found. This was from the 2nd century AD right up through Anglo-Saxon/Scandinavian periods into medieval times. Cores, pips and stalks. Often found in wells, ditches and latrines. Even one complete apple!

York was founded in AD 71 and was a crossing place on the River Ouse as well as a port. Roman crab apple remains were found in the legionary fortress which was garrisoned by 6,000 soldiers. That is a lot of apples to keep the legionaries happy. Apple remains were also found in a Roman well in the Bedern, a very ancient part of the city which in post-medieval times was the red-light district. More apple remains have been found in Skeldergate and Rougier Street. Even imported figs. The Anglo-Scandinavian or Viking period also has many remains, mainly found in old latrines. The population of Jorvik was around 10,000, quite a customer base. Yorkshire apples these days have names like Ribston Pippin, Yorkshire Beauty, Flower of the Town, Cockpit Improved, Yorkshire Greening, Hornsea Herring, Helmsley Market and Fillingham Pippin. As a walled city York is full of old gates: maybe there should have been an Applegate?

Celtic apple legends

Celtic apple myths are very strong and diverse. Suffice it to say that in Ireland there is a famous epic tale, *The Voyage of Bran*, where we see the full power of apples and Celtic women who exert their influence over men. Here the timeless Irish hero Bran mac Febail, a man with a terrible thirst for knowledge, embarks on a voyage, just like Odysseus. While Bran is out walking, he hears beautiful music, as you do in Ireland, and, because the music is so lilting, beautiful and mournful, Bran assumes it comes from uilleann pipes. But no matter where he looks he cannot see the musician. He lies down and is lulled to sleep by the beautiful music. There, in the sacred mists of his mind, Bran glimpses the other world, a supernatural realm of everlasting youth, beauty, health and joy. Not bad for a start. Upon waking, Bran sees beside him a beautiful silver branch with white blossoms, a clue that he is on the right track. It is the 'silver branch of the sacred apple tree' which symbolizes entry into the 'other' world.

Bran returns to his royal house, carrying the silver branch. While his royal company is there, a fine woman from the other world mysteriously appears, and sings to him a poem about the land where the branch had grown. In the song, she tells him that the branch has come from an apple tree in the land of *Emain Ablach*, which may be the Isle of Arran or the Isle of Man.

Like the sibyl carrying a bough who escorts Aeneas down into the underworld in Virgil's epic poem *The Aeneid*, the lady from the Irish 'other world' tells Bran what to expect. It is always summer, there is no want of food or water, no sickness or despair ever touches the perfect people, and there is no unhappiness and 'no rough or harsh voice'. There is an 'ancient tree with blossoms, on which birds call to the hours.'

There is also talk of sweet music and the best wine (or is it cider?). The good lady has fifty verses which she sings in Old Irish and she encourages Bran to visit the Land of Women across the sea. The silver apple blossom is the navigation aid to an Irish Eden where women rule supreme in beauty and knowledge. Being wise, the woman from the other world takes back the silver branch from his hand and vanishes into thin air.

Fired up by the thought of the island, Bran gathers nine men about him, each of whom chooses three men so they have a strong crew and set sail. After only two days and two nights Bran meets the sea god, riding in a chariot, who tells him that he will beget a great warrior once he is back in Ireland. There is talk of sea horses and rivers pouring forth honey, of speckled salmon, calves and lambs and pastures laden with flowers. There is also talk of serpents and sin, greed, lust and vigorous bedfellows. Fairy knolls, dragons and wolves. The sea god has thirty verses of his own. Then they part.

Bran then reaches the Isle of Joy – where everyone laughs, but they all fail to answer a single question. They are too busy laughing. Bran sends a man ashore who is bewitched by the islanders and becomes just like them and starts laughing. A joker. So he is left behind. One is reminded of *Gulliver's Travels* visiting strange worlds . . .

Next stop the Land of Women where they are hesitant about going ashore, not sure what lies in store for them. Noticing their uncertainty, the leader of the women throws a magical ball of yarn across the water towards the boat which sticks to Bran's hand as he picks it up. Being as strong as all the men, she pulls the boat ashore single-handed. And just like a Celtic *Love Island*, each man pairs off with a woman, Bran with the leader.

The men lose all track of time and after what they think is only a year they feel homesick for Ireland. Although warned by the leader of the women not to do so, they set sail and pick up the man they left behind on the Island of Joy. They return to Ireland but when they get there the people who greet them do not recognize them. They have been away for so long that their names are known only in legend. One man jumps ashore and turns to ash; the others, fearing the same fate, simply tell their story, which is carved on stones in Ogham. These are tossed ashore, which is how we know their story today. Bran and his men set sail, never to be seen again. These Celtic women were strong and uninhibited, both in myth and reality. Apple blossom worked its magic.

Then there is another legendary Irish seafarer called Mael Duin, the son of warrior chieftain named Ailill Ochair Aghra. His mother was a nun who was raped by Ailill. His father was then killed by marauders and Mael Duin, after being adopted by a king and queen, later sets out in a boat with his stalwart crew to find the marauders to avenge his father's death. Like Gulliver, he finds many islands, nearly thirty in all, each with a different world or characteristic. One island has ants that eat their boats, another has a woman who pelts them with nuts. Then there is the island with a river and a sky that rains salmon. Others have black and white sheep that change colour, maidens and intoxicating drink, more cattle, more maidens and eternal laughter. Another island has a colourful bird singing psalms and yet another has a golden wall around it. Shades of Paradise?

There is an island with the branch of an apple tree, where they are all fed with apples for forty nights. A land of plenty. There are several variations on this story, one of which involves grafting, where Mael Duin, the budding Irish pomologist, cuts himself a rod which grows into an apple tree bearing three apples which sustain him and his men. Fertility, magic and boundless food.

They find an old man in the sea near Tory Island who has been cast out as a punishment. The man tells Mael Duin than when he meets the marauders, his father's murderers, as he surely will, he should not kill them but make peace. Another Irish Odysseus well on his way to redemption.

Irish Kings, Caledonia and Wales

Many other Celtic myths also involve apples, royalty and salmon. In the Ulster Cycle of Irish mythology, Cú Roí, King of Munster and 'hound of the battlefield', is killed by his rival Cú Chulainn who took his wife Bláthnat – 'Little Flower'. The soul of Cú Roí was confined in an apple that lay in the stomach of a salmon which appeared once every seven years. Cú Chulainn managed to escape by following the path of a rolling apple.

Then there is the Tragedy of Bailé the Prince of Ulster who was riding south to meet Aillin, his betrothed, Aillin. On the way Bailé met an old man who said that Aillin was dead and so overcome was the Prince that he died on the spot. His followers continue south and tell Aillin of their predicament. She also is overcome and dies on the spot.

An apple tree was said to grown out of her grave and a yew, tree from his. Apples and yew, the everlasting green tree. Symbolic and fitting.

In the Irish tale 'The Adventure of Conle', Conle the son of Conn is fed an apple by a fairy lover, which fully sustains him with food and drink for a month; however, it makes him long for the beautiful country of women to which his lover is enticing him. In the Irish story from *Oidheadh Chlainne Tuireann*, the first task given the Children of Tuireann is to retrieve the Apples of the Hesperides (or Hisbernia). So maybe there is a Greek maritime link after all.

Here again are the same elements: beautiful women, apples, 'trouble and strife'. Apple wood was also sacred and burned by Celts during fertility rites and festivals. Such apple beliefs go back much further than Christianity. In pre-Christian Brittany, for example, according to the French food writer Maguelonne Toussaint-Samat apples were eaten before the prophesies were made, and the magician Merlin sat under an apple tree to teach. The

Breton St Konorin was reborn by means of an apple. Gardens, apples, orchards and education are one and the same. The trees are our books, the apples our inheritance, the leaves our thoughts.

In Wales there are also many apple stories. 'Afallenau' ('Apple Trees') is a twelfth-century Welsh narrative poem dealing with Myrddin Wyllt. And in *The Black Book of Camarthen*, apple trees are worshipped and eulogized:

> Sweet apple tree that luxuriously grows!
> Food I used to take at its base to please a fair maid . . .
>
> Sweet apple tree, and a tree of crimson hue
> Which grew in concealment in the wood of Celyddon

In Wales, as in Ireland, apple myths and old stories are a vital part of the living culture. Cider making in Wales almost completely died out after the Second World War, though there are those who remember it being drunk on Welsh farms particularly at hay-making time. There was even a cider pub in Cardiff Docks for thirsty dockers. Morgan Sweet apples were taken across from Somerset for the South Wales coal miners. Monmouthshire was a prime cider county and has good orchards and old perry pear trees. Luckily, Welsh cider is enjoying a revival and enthusiasts are hunting down old Welsh varieties of apple, including the windswept Bardsey apple *Afal Enlli*, found growing against a wall on Bardsey Island in North Wales. It may be descended from a variety planted by monks as far back as the thirteenth century. The abbey there was founded by the Welsh kings and for centuries was a place of pilgrimage. Twenty thousand saints are supposedly buried there, the bravest and best in the land. There was also an orchard. An island of apples and saints. The Bardsey apple is free from scab and canker which means it is disease-resistant. Must be the salty air!

The Celts also believed that apples were associated not just with fertility but rebirth. Apples were said to be buried in graves as food for the dead on their final journey. Petrified remains of sliced apples have been found in Western Asia in tombs dating back to 5000 BC.

In the Neolithic village of Skara Brae in Orkney archaeologists have also found remains of crab apple pips dating back to about 3000 BC. Skara Brae is remarkable. Crab apples may have grown locally or been traded. The inhabitants had large pottery vessels known as 'grooved ware' used for storage or fermentation. Where there's a will, there's a way. If you were making cider even in small quantities you needed a secure vessel for the fermentation. Wood was scarce. As always, good, safe storage is the key.

In Scotland there are several ancient apple customs and myths including *Lamaesabhal* or *Lamas Ubhal* an early pagan Celtic festival which gives its name to the drink associated with it, which can be hot cider or ale with spices and roasted crab apples floating on top. It is drunk at Halloween in a wassail bowl, either wooden or ceramic. In Celtic mythology, Lugh, one of the prominent gods, is portrayed holding three apples, symbols of immortality, power and prosperity. No wonder cider had such a hold on their imagination. It was in Scotland that the wife of

Argentocoxus, a local chieftain, once waxed lyrical to the Empress Julia Domna about the freedom of sexual relations in Celtic society. As the chieftain's wife told her, 'We fulfil the demands of nature in a much better way than do you Roman women; for we consort openly with the best men, whereas you let yourselves be debauched in secret by the vilest.' *Touché*. Julia was the wife of the Roman Emperor Septimus Severus. Cassius Dio reports the conversation in his history, but sadly no mention of cider. In England this drink is called lambswool and is often made with cider and drunk on All Saints Day, Christmas eve, winter solstice or twelfth night. The ancient pagan ritual well preserved.

Evidence for Scottish cider making is difficult to find. Records of archaeological remains of apples in Scotland are scant and it would seem that cultivation only started in the Norman period and was associated with monasteries and abbeys. Charred plant remains have, however, been found on Mesolithic sites in Scotland and have included crab apple and pear pips, hawberry pips, hazelnut shells and the roots of lesser celandine. Twenty-one fragments of wild apple were recorded at a Mesolithic site on the island of Colonsay (along with 30,000 - 40,000 hazelnuts). Wild crab apple trees have been found in many places in Scotland: for example, Dumfries, the Central Belt, Loch Lomond, West Coast, Perthshire, Aberdeenshire and a few in the northern highlands. The sight of a large, old crab apple tree in full blossom in a remote mountain forest is spectacular.

Apples also grow well in the Carse of Gowrie between

Perth and Dundee with names like Lass O'Gowrie, Galloway Pippin, Scotch Bridget and Bloody Ploughman. There has recently been a revival in cider making in East Lothian and Dumfriesshire using home-grown apples. Scottish cider is no longer a myth but a reality.

Ancient Britain

Celtic Britons knew all about indigenous crab apples which often grow in strange, wild places and give an amazing display of blossom each year. These ancient Britons also knew about oaks, sacred groves, hunting the wren, as well as druids, woad and mistletoe. Shades of wassail. Perhaps one day archaeologists will find out that Stonehenge was built as part of a midsummer cider festival where people congregated from all over the country to judge and then drink the cider. They had a rich and vivid imagination. Later apple myths ran in parallel with those of ancient Greece and Rome. There were strong connections through tin trading with the Mediterranean as well as the Phoenician and Greek colonies.

Greek geographers knew all about ancient Britain and Ireland. The first known Greek visitor was Pytheas of Massalia, or Marseilles (*c.*325 BC), who visited Great Britain and Ireland. He may have got as far as Iceland as he describes the Arctic, polar ice and the midnight sun. He mentions tides and their relation to the moon and coined the word *Thule*. The furthest north location referred to in classical times and later on in medieval times as *Ultima*

Thule - a mythical island - the borders of the known world.

Sadly, none of his writing survives, but his observations are quoted by Diodorus Siculus, Timaeus, Posidonius and Pliny the Elder. Pytheas mentions tin being traded in small ingots like knuckle bones, ingots that were transported at low tide in wagons to the island of Ictis, which experts believe may be St Michael's Mount in Cornwall. Diodorus also says that the inhabitants of Cornwall were civilized in manner and especially hospitable to strangers because of their dealings with foreign merchants. Tin extraction is thirsty work even above ground. Sadly, there is no hard evidence that they made cider at this stage. If there was, they must have drunk the evidence. But crab apples would have existed.

Cross-Channel trade was also common in iron age times. There was easy access to Brittany, Normandy and the Basque region. Grafts could be transported wedged in lumps of clay across the Channel and along the Celtic seaboard from Spain. Strabo (64 BC–AD 24) is believed by many to have commented on cider, or *sidra*, being made in northern Spain in Asturias, but there is no real evidence for this in literature or that he even went there. The actual text reads:

> χρῶνται δὲ καὶ **ζύθει**: οἴνῳ δὲ σπανίζονται, τὸν δὲ γινόμενον ταχὺ ἀναλίσκουσι κατευωχούμενοι μετὰ τῶν συγγενῶν

> They also use beer; wine is very scarce, and what is made they speedily consume in feasting with their relatives.

The key word is 'zythos', which is also mentioned by Strabo as a drink of the Alexandrians in Egypt and is unlikely to be cider.

But the notion of mythical orchards overseas, like offshore bank accounts, was the order of the day, as if over the horizon there was always a promised island beckoning. But some of these apple islands were much nearer to home. Right on our doorstep in the very heart of cider country, in fact. Apple islands surrounded by marshes, wetlands and inland seas.

One of the most enduring of these myths is all about Avalon, the mythical Isle of Apples where healing comes in many shapes and forms. In Latin it is known as *Insula Avallonis*, in Welsh *Ynys Afallon* and in Cornish *Enys Avallow*, which means island of fruit or apple trees. It is also the legendary island that featured in the Arthurian legends first written down by Geoffrey of Monmouth in 1136 and very conveniently linked to Glastonbury Abbey. Geoffrey claimed that the name came from Avalloc, who once lived on this island with his daughters.

It is Geoffrey's story about Arthur in *Historia Regum Britanniae* (*The History of the Kings of Britain*) that caught the medieval imagination. It was in part derived from an earlier account by Nennius, a ninth-century Welsh historian well versed in Latin.Avalon was the place where King Arthur's sword Excalibur was forged and later where Arthur was taken to recover having been gravely wounded at the Battle of Camlann when he was fighting Mordred. Arthur, mortally wounded, is carried on barge with damsels to the Isle of Apples. No doubt mulled cider featured in his treatment. There are many fine orchards around Glastonbury to this day.

The Isle of Apples was also called the Fortunate Isle (*Insula Pomorum quae Fortunata uocatur*) because it produced all manner of fruits on its own accord, just like the wild apple forests of Kazakhstan. This also applied to crops, whose fields needed no ploughing and where the only cultivation required was by nature itself. It produced grain and grapes, and apple trees grew in its woods from the close-clipped grass. People lived for a hundred years or more. There nine sisters ruled by a pleasing set of laws over those who came to them from our country.

The district is also known as *Ynys Gutrin* in Welsh – the Island of Glass – and from these words the invading Saxons later coined the place-name 'Glastingebury'. Other writers took up the theme, notably William of Malmesbury. But it was Sir Thomas Malory (1405–71) in his epic tale *Le Morte d'Arthur*, written while in prison, that secured Arthur's reputation forever. Arthurian legends were now here to stay. It was these legends which, 500 year later, drew the American novelist John Steinbeck to Somerset in the late 1950s. Legends which some have speculated underpinned plots in his novels. He spent a year researching the Arthurian legends and lived near Bruton. Steinbeck first read *Le Morte D'Arthur*, when he was a 10-year-old boy.

Even as late as 1586 William Camden, the Elizabethan antiquary, takes up the theme of Avalon with gusto:

The Isle of Apples, truly Fortunate
Where unforc'd goods and willing comforts meet.
Not there the fields require the rustick's hand,
But Nature only cultivates the Land,
The fertile plains with corn and herbs are proud
And golden apples smile in every wood.

The enduring themes are very clear to see: apples and young women, symbols of youth, fertility and eternal life. As well as voyages both real or symbolic, akin to the journey of the apple and life itself. Maybe all these elements of voyaging combine today with the extraordinary success of the iconic, long-running midsummer pagan event known as the Glastonbury Festival, or just 'Glastonbury'. It has been going since the 1970s, initially organized by Michael Eavis and latterly by his daughter Emily. Here you will find incredible music and oceans of cider, as well as many colourful side events, including fire-eating, juggling, women's mud wrestling and storytelling.

One night I helped serve cider there from the famous Burrow Hill Cider Bus, close to the main pyramid stage when spectators were ten or twenty deep. That was quite an experience. Some customers even wore Viking horns.

Many of the eco-experiments demonstrated in the green field thirty years ago, such as solar panels, have now gone mainstream. I also gave a demonstration of wind-powered sheep shearing, which alas is not mainstream.

On the air waves in the early days there was Radio Avalon, a pirate radio station now called Worthy FM. Glastonbury Tor, the Isle of Apples, can be seen only a few miles away, though 500 years ago the abbey's cider would be taken from their orchards in the festival village of Pilton by boat. So Arthur could easily have hitched a ride with the apple pickers. Myths, legends, Aga sagas, apples and cider are deeply ingrained in this part of Somerset. Eternal youth or just healthy old age?

4

MEDIEVAL CIDER

The Romans left these shores in the fifth century AD and returned to Rome. The Romano-British estates soldiered on for a few generations. This long period of uncertainty is known as the Dark Ages. Roman cities like Bath decayed. 'Age underate them' is how the unknown earliest poet of 'The Ruin' describes this once glorious city. People became more nomadic and pastoral in order to survive. Did King Arthur and his father Uther Pendragon make cider? We shall probably never know, but there were strong links between the West Country and Brittany for several hundred years.

What then followed became known as the Middle Ages and lasted till the late fifteenth century, roughly a thousand years of medieval England. The Saxons, it seems, preferred beer and ale. There was apple wine, called *æppelwīn*, mixed with honey to give it a sweeter kick. Scholars still debate the extent of Saxon cider making, but by the tenth century Saxon abbeys and monasteries had orchards so it was perfectly possible.

Everyone remembers 1066 and all that. William the Conqueror defeating the Saxons and claiming England for himself. Battle of Hastings. Vikings by the back door. One in the eye for Harold. The upshot was that the Normans brought with them heavy-duty, Roman-style cider-making equipment and cider expertise. This equipment had been used in France for nearly a thousand years and now it was about to cross the Channel. This basic but all-important technology had evolved from two sources: wine making and olive oil extraction. The Greeks and Romans loved their wine and olive oil. They needed vast quantities. In Rome it was estimated that the consumption of olive oil was around 25 litres per person per year, double what it is today. Today: 20 litres in Greece, followed by Spain 14, Italy 11 and Portugal 8 litres.

Ancient classical civilizations faced all the same problems of growing, picking, processing, storage and warehousing, transport and distribution that artisan cider makers do today. Let alone dealing with middlemen and advertising. But human ingenuity knows no bounds where food and drink are concerned; efficiency and optimization without compromising quality are top of the list. A maxim that could be applied to artisan cider making today. Methods that the Romans evolved were used in England for over 800 years.

Pressing olives is hard work and often a two-stage process, first crushing olives in a mill to break up their structure then pressing them to get the oil. Just like apples in fact. The Mediterranean is still littered with circular olive mills with edge stones like vertical millstones running round a circular trough. This would have been powered by horse or donkey. Anyone making cider in England a hundred years ago would instantly recognize these mills without realizing their ancient pedigree. I have seen circular mills still used in Cornwall, Devon, Somerset, Herefordshire and Jersey.

Olives were first thought to have been crushed in mats made from fibre with large stones placed on top. As old as the hills. It was basic but it worked. Then, around 400 BC, a Greek philosopher, mathematician and astronomer called Archytas of Tarentum (428–350 BC) invented the screw thread. This technology was subsequently applied to wine and olive pressing. Threads were carved out of solid wood. A great skill. Again, the same as can be found in Devon, Somerset, Herefordshire or Jersey. Rustic cider – *méthode traditionelle*.

Sometimes the Romans used beam presses to extract

Fig. 4.1 'Ancient olive press, with windlass', Arboretum Tresteno, Croatia', 17th century

olive oil. The beams were enormous, often entire tree trunks. To get extra pressure windlasses were used to tighten a rope around a vertical wooden axle with a handle. Basic engineering. Levers and mathematical advantage. Archimedes again.

This is basic olive oil-pressing technology that was brought across the English Channel by the Normans in the twelfth century and was put to use in monasteries for crushing apples to make cider. Some farms still use it today. At Hellens, a wonderful house in Much Marcle with a drive lined with perry pear trees, I have seen strong men putting their shoul-

Fig. 4.2 'Ancient & Modern', Clifford Lugg and his son Barry making cider at Tregarne Farm, St Keverne, Lizard, Cornwall, 1978

ders to the wheel and crushing the perry pears. Beam presses are less common but there is one small one in Cornwall still in action at St Keverne on The Lizard as recently as 1978.

These men of The Lizard are far closer to the delights of Roman wine making and olive technology than they perhaps realize. Old ways are often the most satisfying. Where there is a will, there is a way.

St Jerome and cider

But all this raises a number of questions. Was cider ever made in Britain before the Romans got here? If so, how was it made and was it drinkable? And did it continue to be made after the Romans left?

One persistent rumour down the ages and oft repeated in many cider articles and websites is that Julius Caesar came to Kent and found the ancient Britons drinking cider. This story still circulates today though it has no foundation in Caesar's *Commentaries*. Caesar invaded in August 55 BC with two legions and again in 54 BC with 628 ships, 5 legions and 2,000 cavalry. Both times in Kent. They timed it right for the dessert apple harvest. But where was the cider? And if the Kentish cider was any good, why did the Romans wait nearly ninety years till AD 43, when Emperor Claudius mounted a major campaign to subdue the Celtic tribes?

I decided to enlist the help of a Classics expert, Bijan Omrani. If anyone would know about Caesar and cider I felt sure Bijan would have the relevant passages at his fingertips. I waited a few days then his e-mail appeared:

> I have been turning over my books on the question of Caesar and cider. I'm sorry to say that the whole story appears to be the most extraordinary canard! I see that a number of writers have quoted the story of Caesar reporting cider-making during his invasions, but nowhere is a citation given from the primary text.
>
> I have been through Caesar's Gallic Wars, and I can find no mention at all of Caesar or his troops observing cider-making in Britain, or anywhere. Indeed, the word 'sicera' does not appear in Caesar, and I can find no mention of apples either.

The word 'sicera' seems only to appear in late Latin, and was used several times by St Jerome (b. AD 347) who takes the word into Latin from Hebrew. Jerome was a scholar of Hebrew and Greek and best known for his translation of the Bible into Latin – the Vulgate Bible.

Jerome's understanding of *sicera* is given clearly in Epistle 52:

> Priests given to wine are both condemned by the apostle [1 Timothy 3:3] and forbidden by the old Law. Those who serve the altar, we are told, must drink neither wine nor *shechar*. Now every intoxicating drink is in Hebrew called *shechar* [*sicera*] whether it is made of grain or juice of apples, whether you distil from the honeycomb a rude kind of mead or make a liquor by squeezing dates or strain a thick syrup from a decoction of grain.

So the juice of apples was writ large:

> Whatever intoxicates and disturbs the balance of the mind avoid as you would wine. I do not say that we are to condemn what is a creature of God. The Lord Himself was called a wine-bibber and wine in moderation was allowed to Timothy because of his weak stomach.

Bijan Omrani had sorted out an age-old riddle. Cider was obviously not of great importance to St Jerome, though he advocates a clear mind. Always useful when cider making.

Dark Ages and St Teilo

In AD 410, only ten years before Jerome's death, the Romans left Britain. The Roman Empire was collapsing thanks to the attentions of Goths, Huns and Vandals, Visigoths and Ostrogoths. What happened to cider making is unknown, but apple trees still grew and blossomed. Over the years Christianity took root and grafted itself onto Celtic culture, often in Wales and Ireland, with monks living in outlying islands and caves. Knowledge and learning was preserved within the four walls of monastic cells that later became monasteries.

In AD 513 according to the Bishop of Poitiers, Queen Radegund, wife of Clothar, King of the Franks, often fasted and drank nothing but mead and perry. The height of fashion. She became a Saint. In the sixth century AD Brittany was close to England politically and a good place for apples.

After the Yellow Plague which swept through Wales, St Teilo took the survivors of his community to Dol in Brittany where he and St Samson planted groves of fruit trees several miles long from Dol to Cai. There are still fruit trees there today named after Teilo and Samson.

The Yellow Plague, or Justinian Plague, came from the Tien Shan around AD 541–42, with recurrences until 750. It was a forerunner of the Black Death and killed off half of Europe. Kazakhstan and Khirghizstan are where apples come from. A strange coincidence. It may have been brought inadvertently by nomads or Asiatic rat fleas.

In 664 there was another Yellow Plague in Britain which coincided with a solar eclipse and lasted twenty-five years. Our sins must have been very great. Settled agriculture reverted to a more nomadic pastoral existence. Maintaining orchards and vineyards was difficult. They tied you down to one place, so it was better to be mobile. In the eighth century St Ségolène, Abbess of Troclar, refused all drinks during Lent except water and perry. Penance and abstinence. Without perry there was no God. Or maybe God was pear-shaped? Apples were always close to Christianity. There is even an eighteenth-century carol entitled 'Jesus Christ the Apple Tree'.

When researching a talk for the Abergavenny Food Festival I came across St Teilo once again, who ought to be the patron saint of Welsh cider. The vast two-volume tome *Herefordshire Pomona* of 1878 had this to say about Welsh cider: 'In 1612 more cider was produced in Monmouthshire than in Herefordshire . . . The tithes for Llantilio Crossenny not seven miles from Abergavenny were 50 hogsheads.' That meant that the overall production of that one village was 500 hogsheads, or 27,000 gallons, roughly equivalent to 120 tons. The church was dedicated to St Teilo, hence the name of the village.

After the talk, a man came up to me with a very broad smile on his face and simply said, 'I am the vicar of Llantilio Crossenny – can we reinstate tithes?' I don't blame him. The first reference to cider in the Welsh language dates from the second half of the fourteenth century and references to cider appear in Welsh poetry from then on.

Cider was very popular in southeast and mid-Wales. Like Herefordshire, very good soil and climate for apples. At one time almost every large farm had its own orchard and cider mill. Welsh varieties of apples and pears are often distinct from those grown in England, giving Welsh cider a different flavour. There are many old granite or sandstone circular troughs housed in barns to this day.

The Dark Ages are indeed a little dark, what with plagues, famines, bad weather and invasions by Saxons, Vikings and Danes. Literature and cider making took a hammering. A black hole in other words, though religious houses like Ely had orchards in Saxon times. There was hope.

Charlemagne's edict

As the Roman Empire imploded and split in two, there was a power vacuum. The western Roman Empire collapsed in 476, but the eastern imperial court survived another thousand years until the fall of Constantinople in 1453. Britain was left out on a limb to fend for itself. Independence of a sort. A wild native 'Celtic' land which welcomed Christianity and communion. Cider making was on the back burner.

Charlemagne (748–814) was crowned Holy Roman Emperor by the Pope in AD 800. So Roman ideas and laws were reinvented and reinterpreted. The Holy Roman Empire had arrived and Charlemagne took up the reins. Monastic houses flourished. Economic and political stability allowed orchards and vineyards to prosper. He was keen to administer his realms efficiently. Edicts were promulgated outlining how royal estates should be run as well as stipulating the extensive duties of stewards. The *Capitulare de Villis* was composed *c.*771–800. It was a blueprint of estate management, common sense, fairness and accountability, a deep insight into what mattered 1,200 years ago and all of it is appropriate today.

Wine, grapes, vineyards and wine presses were near the top of the list (for *wine*, substitute the word cider):

> Our stewards shall take charge of our vineyards in their districts, and see that they are properly worked; and let them put the wine into good vessels, and take particular care that no loss is incurred in shipping it. They are to have purchased other, more special, wine to supply the royal estates. And if they should buy more of this wine than is necessary for supplying our estates they should inform us of this, so that we can tell them what we wish to be done with it. They shall also have slips from our vineyards sent for our use. Such rents from our estates as are paid in wine they shall send to our cellars.

. . . but herbs, vegetables, apples and other fruits appear near the bottom. Charlemagne was keen on his fruit and concubines.

> As for trees, it is our wish that they shall have various kinds of apple, pear, plum, sorb, medlar, chestnut and peach; quince, hazel, almond, mulberry, laurel, pine, fig, nut and cherry trees of various kinds. The names of apples are: *gozmaringa, geroldinga, crevedella, spirauca*; there are sweet ones, bitter ones, those that keep well, those that are to be eaten straightaway, and early ones. Of pears they are to have three or four kinds, those that keep well, sweet ones, cooking pears and the late-ripening ones.

The seeds for good estate management and vineyard practice had been sown. Ideas about good husbandry espoused in the edict involving wine making were applied to orchards, apples and cider making on large estates and in a monastic context.

Domesday Book

The Normans invaded England in 1066. But who were they? Vikings in disguise? In the wake of Charlemagne's death in 814, the Vikings decided to occupy Normandy, which they had been raiding. In 841 their fleet turned up at the mouth of the Seine under Hrólfr Ragnvaldsson, known as Rollo. After sacking a few monasteries, they settled down and became a vassal state under Charles the Simple.

Over the next 100 years the Vikings went native, learned Norman French and how to make and drink Normandy cider. They ran a protection racket. By 1066 they had itchy feet. William the Bastard was a descendant of Rollo and his fighting gene needed at outlet. The Normans invaded England, won the Battle of Hastings and drank Sussex cider. The rest is history. William's father was Robert, Duke of Normandy, and his mother Arlette helped found a Benedictine monastery. Grestain Abbey was very influential in taking over the Church of England and founded many churches and priories. William the Bastard now became William the Conqueror. Architecture changed from Saxon to Norman. Beer to cider.

Like Charlemagne, William was keen on his estates and wanted to know how much England was worth, so he ordered the creation of Doomsday Book. A chance to see how many orchards there were? Sadly, not a huge number are mentioned, but Pillocks Orchard in Somerset, Stoke Orchard in Gloucestershire and Cann Orchard in Cornwall all get a name check.

Vineyards are more plentiful. There are thirty-two in the southeast of England and many were recently planted. Orchards were small and of no great interest. It was when vines began to fail that orchards took over.

William the Conqueror died on campaign back in Normandy only a year after Domesday Book was completed. So he didn't have much time to read it.

Medieval orchards

It was only after a century of quelling rebellious Saxons that English cider making got into its stride. Monastic records prove this. With access to cider presses and apple mills, monasteries became important as monks started to grasp the basics of science and mechanics. Their desire to drink good wine and good cider was the key. They used their intellect to devise their own cider philosophy.

The Rule of St Benedict allowed a modest ration of wine per day and more on feast days and holy days. Set out in AD 516 by Benedict of Nursia, each monk was allowed a pound of bread and a *hemina* of wine (a Roman measurement equivalent to about a quarter of a litre). The monk had to read an edifying book in the evening, then observe strict silence after compline. He must sleep in his habit so that he was ready to jump up for prayers early in the morning. Regarded as last year's apple juice, cider was not rationed within the Rule, hence its popularity.

The path to high-quality wine and cider was established in the minds of the monks who became more specialized in chemistry, alchemy and scientific knowledge that helped their craft. Many religious houses in England were connected to mother houses in France. Such knowledge was often fiercely guarded even within their own tight-knit communities and only shared with those they could trust. The climate changed in the thirteenth century when vines became less productive and orchards took over: apples were more reliable.

In early references orchards in a religious context are sometimes called *pomeriums*. The word has nothing to do with apples, but is a corruption of the phrase '*post moerium*', literally 'beyond the wall', and goes back to Etruscan times. *Pomerium* was originally used to define a religious space, either a small strip of land or a boundary mark made by a ploughshare or wall around a sacred building. It was forbidden to carry arms within the area or bury dead within the *pomerium*. But by medieval times, when orchards were becoming popular, it was customary for monks to be buried in them. The sense of the orchard as a sacred place is deep-rooted in early Christianity.

Orchards were also referred to as *hortus*, or *ort*, Latin for gardens. The suffix *-geard* means 'fence' or 'fenced-off land', which became yard, as in 'backyard', 'farmyard' or 'churchyard'. *Ortgeard* or *orceard* is derived from *Hortus-geard* which becomes 'orchard'. *Hortus* also lends itself to horticulture. Sometimes gardens and orchards are quoted separately in documents, sometimes not.

Reading through medieval records, eagle-eyed archivists familiar with various shorthand notes and strange writing styles can detect orchards at a distance of nearly 1,000 years. There are over 300 references to orchards during this period. Examples of *pomeriums* or *orto*, *ortgeards* or *hortus* can be found for religious establishments at Ely, Thorney, Canterbury, Kirkstead, Dunster, Coventry, Gloucester and Bristol.

William of Malmesbury (1095–1143) talks about the Vale of Gloucester, where 'The apple and pear trees of the district yield most profusely liquors which in the estimation of the inhabitants are far preferable to any wine from foreign countries.'

What is fascinating are medieval references to cider making which crop up in monastic and port records. They can reveal quite a bit. Remarkably, many of the places mentioned remain key cider areas today.

Monastic cider

Monasteries became the powerhouses behind cider making. They became very wealthy and acquired more land. They had capital to build cider presses within tithe barns. They owned orchards and received cider as rent and tithes. They also had a ready workforce and a captive clientele. If the grape harvest failed, then *cidre* was the next best thing. Existing monastic records can still be consulted. These often appear in the form of rents, purchases or cider sales. Cider presses were rented out. It was a very fluid economy.

Luckily there are clues in the archives. The historian and monk Gerald of Wales (*c.*1146–1223) notes that the monks of Christ Church, Canterbury, preferred cyder to Kentish Ale and that cyder was drunk alongside claret and mulberry wine. These monks were well known for their sixteen-course dinners. Such decadence was in the end the monasteries' undoing. Henry VIII became jealous.

What is interesting is that cider records are scattered up and down the country, from Kent to Worcestershire, Devon to Yorkshire. One of the earliest surviving accounts is from Staffordshire in the year 1200 at Loxley, near Uttoxeter, of 'le Pressurhus', with outbuildings including a cider press (*molendina ad poma*) and land purchased with a meadow thrown in for good measure.

Then, four years later, the tenancy of Runham Manor in Norfolk, was paid for with cider – two *mues* of 'wine made from pearmains', a *mue* being a vessel or container. So in King Edward I's time 'pearmain cider' was called wine. These records are invaluable because they not only give the name of the tenant but the value of cider. In Norfolk there is also an orchard in Banham producing three casks of cider annually, each worth 10*s*. And in 1205 Robert de Evernue held his Lordship of Redham and Stokesley by paying 200 Pearmains and two hogsheads of cider. Not a bad deal. What may seem like a boring list becomes fascinating as you piece together the places and people. It is rather like a vast cider jigsaw.

Pearmain is an interesting name for apples. Some say it means 'great pear', others that it comes from the Old French *parmaindre*, meaning it will 'endure', i.e. it is hard and keeps through the winter, hence 'Winter Pearmain'. This is the earliest named apple in England, not to be confused with the Victorian apples Worcester Pearmain and Adam's Pearmain.

Many references to cider can be found in the southern counties of England closest to Normandy. The Normans liked to have a foot in both cider camps and many religious houses did just that. In 1212 cider is recorded as income for Battle Abbey, Sussex. The abbey was founded a few years after the famous battle. The story goes that

Pope Alexander ordered the Normans to do penance for killing so many Saxons during their conquest of England. William the Conqueror, never one to miss a PR opportunity, vowed to build an abbey on the site of the Battle of Hastings, with the high altar of its church on the spot where King Harold supposedly fell. Cider sealed the deal. Later they had a horse-powered cider press which they rented out.

Another large Benedictine Abbey in Normandy at Montebourg held the priory of Loders, near Bridport, founded in the reign of Henry I (1068–1135), a lovely, sheltered valley for orchards which reputedly saw the first cider making in Dorset. Connected to Loders along the coast was the manor of Axmouth in Devon which was making cider in 1265.

Over the border in Somerset cider presses were a source of income in a Royal Charter granted in 1230 to Jocelin, Bishop of Bath. Glastonbury Abbey also had vast orchards and a fine tithe barn which still stands. I have seen cider made there several years running. There is a small orchard around the back with old apple varieties. Then in Glastonbury Brother Robert bought mulling spices – 3 lb of ginger, honey, cloves, mace and rock sugar – no doubt anticipating Michael Eavis and the Glastonbury Festival. Mulled cider always goes down well if it is raining.

But even in medieval times trees were expensive and highly valued, particularly if they originated in France. In 1265 the Bishop of Winchester's bailiff bought 129 apple and pear trees for an orchard at Rimpton, near Yeovil. The cider economy was on the up. There are still orchards in Rimpton today, close to the old Magna Cider Co., in Marston Magna, not to be confused with the Irish cider company Magners.

Even the nuns of Shaftesbury Abbey in Dorset owned large orchards and were making cider. Maybe they were the first women cider makers. In 1291 there is mention of cider, *cisera*, in an enrolled account. It was said that if the Abbess of Shaftesbury were to marry the Abbot of Glastonbury they would have more land than the king. Shaftesbury Abbey was founded in 888 by Alfred for his daughter. At one time I lived in the Shaftesbury area opposite a place called Parhams Farm, Saxon for 'pear village' farm. So they may well have had perry as well. Lucky nuns.

What is also interesting about medieval cider is the trade in cider and ideas back and forth across the English Channel. Priories were connected to their mother houses in Normandy. Cider knowledge was exchanged freely with the Continent. In 1265 Beaulieu Abbey in Hampshire sold about 2,400 gallons of cider and this would have been shipped to Southampton or London. Cider ports were very useful. Just along the coast, in 1320 William of Shoreham, poet and vicar of Chart in Kent, reported a Church edict that stated, 'Young children were not allowed to be baptised in cider.' Cider was often better for you than water. It had at least been fermented and cleansed.

Cider was big business in Sussex and in the 1340s: seventy-four out of eighty parishes in Sussex paid their tithes in cider. Sussex cider was also shipped from Shoreham to Calais, which was about to be besieged. The soldiers were a thirsty lot. This time the English won.

Cider not only moved inland but also into Middle English literature. Between 1370 and 1390 William Langland composed *Piers Plowman* on the Malvern Hills. Vision 8, Passus XX, line 410:

> May no sweet wine or cider or precious drink
> Fully wet my whistle or my thirst slake.

Wetting your whistle was an important part of the literary procedure. Though too much cider can affect your 'vision'.

Not far away, Nicholas of Hereford (d.1420) helps in John Wycliffe's translation of the 'Cider Bible'. The relevant passage from Matthew reads: 'For he [John the Baptist] shall be great in the sight of the Lord, and shall drink neither wine nor "sidir".' (i.e. strong drink). So in Hereford strong drink becomes cider. It still is, though not perhaps as strong as it was. The 'Cider Bible' is held in the Chained Library at Hereford Cathedral.

Geoffrey Chaucer also quotes 'This Sampson never cider drank ne wyn' – from 'The Monk's Tale' in *The Canterbury Tales*. Chaucer lived in North Petherton in Somerset when he was Deputy Forester from 1391 to 1400. Cider was everywhere and when the apple crop failed, cider was imported from Normandy; one year French cider was shipped into Winchelsea which often acted as a staging post for London. Once in a barrel or hogshead, cider could be loaded and unloaded either up gangplanks or straight into the hold with a small crane and strops. Once on board a small sailing ship, the cider could go anywhere.All much easier than lumbering along rutted lanes in an ox cart.

Dorset also had quite an export/import trade. In the reign of Edward IV, 'A vessel named the *Mavye of Reyle*, brought into Poole as part of its cargo 1 pipe of "sidre" valued at 3s 4d and Stephen Cressyn, a foreigner, paid thereon 1/2d in customs duty and 2d in subsidy.' Continental trade is nothing new. It has been going on for thousands of years. Port taxes made the customs officer's job greatly sought after.

Another vessel, 'the *Barbary of Rever Jobles*', entered Poole Haven under command of her master, Thomas Viron, possibly a Basque from Bayonne. Among the cargo, besides great stores of apples, pears nuts and other fruits of the earth, were three 'puncheons de perry' containing one cask (dolium) valued at 10*s*. On this consignment of liquor customs duty and sixpence, as his share of subsidy. After disembarking her cargo she loaded up with English goods and returned home, but reappeared in Poole two months later with more apples and nuts. But instead of perry she brought a hogshead of dry wine (*vini non dulcis*), a barrel of verjuice and two butts of Runnay or Roumey wine which contained one cask and one hogshead of dry wine. Cross-Channel trade was rife. Even to and from Spain and Portugal.

Growing up in West Devon, I was fascinated by the

history of Tavistock Abbey. About a hundred years of cider accounts still exist. Every day I used to walk to school past the abbey walls alongside the river. The abbey had orchards even in Saxon times. In 1475 ten pipes of cider made from their orchards in Plymstock were shipped across Plymouth Sound. Ten pipes is about 1,200 gallons. This also included three pipes in lieu of rent for use of their cider press. It must have been a fine day. Then they sailed 12 miles up the River Tamar on the tide to Morwellham Quay and carted the cider a few miles over the hill to the abbey. Quite a cargo.

Navigating those waters is never easy as there are some nasty tide rips and mudbanks further upriver. But they knew what they were doing. Wind and tide wait for no man. Tavistock Abbey also had a still tower for distilling cider and herbal concotions. No doubt they had their own inhouse doctors and apothecaries. But the monks in Tavistock were not just interested in cider presses and distilling. They had another sort of press: a printing press. And in 1516 they printed an extraordinary work: *The Consolation of Philosophy* by Boethius. Very learned monks. Philosophy and cider were rife.

These are just a few of the records that survived the dissolution of the monasteries. Cider in monastic times was a thriving business that underpinned much of the local rural economy, keeping fruiterers and coopers busy. Cider is often mentioned in the same breath as wine, as if it *was* wine. Cider must have been well made and in demand as it travels both ways across the Channel. If you could pay your tithes and rent with cider that was better still. It also kept the monks and nuns happy. Cider for the boys, perry for the girls. Make merry . . . That is as good a place as any to leave medieval cider.

5

ORCHARD RENAISSANCE

Slowly England began to shake off its medieval yoke. After the Wars of the Roses there was a new prosperity and confidence, as well as a distinct change in the way people viewed the natural world. Tudors were adventurous. On the Continent the Renaissance was in full swing, in both the arts and science. Scholars were rediscovering ancient Greek and Roman texts. In Oxford and Cambridge philosophy was flying around as if there was no tomorrow. Ideas were on the move and a new national consciousness was emerging from under the monastic cloak.

It was this new vitality and way of looking at the world that filtered down to orchards and cider making which were about to experience their own renaissance. It was as if the last thousand years had vanished and they were carrying on where the Romans had left off. Only this was a peculiarly English Renaissance.

The monasteries thought they could go on for ever even

though these new ideas challenged many religious beliefs. Their spiritual power was dwindling but their wealth remained. New recruits were becoming harder to find. The Church was God's own multinational company with headquarters in Rome. Monasteries were semi-independent, with their own empires. Charlemagne would have admired the expansion of cider making that followed his edicts.

Orchards and ornamental gardens were soon all the rage for royal palaces and aristocrats. Men travelled to the Continent bringing back new varieties of fruit to impress their friends. Orchards can be stunning. What is more beautiful and comforting than an orchard in spring? Bursting forth with blossom and the sound of countless bees going about their diligent work from village to village.

What is more magnificent than an orchard heavily laden in autumn? A harvest of colour that uplifts both young and old with a common purpose and every year inspires artists, composers and poets. The riches of the Renaissance were beginning to bear fruit both intellectually, artistically and scientifically. A thirst for knowledge that had been fermenting secretly for years broke free from its religious chains. People's minds were beginning to open up to the natural world. Was it God's universe or was it theirs to explore?

But while monasteries were ramping up their cider production there had been some radical thinking behind the scenes by two medieval philosophers. Both came from Somerset. Adelard (*c.*1080–1150) lived in Bath Priory. He translated Euclid and many other mathematical texts via a circuitous route. Many of the Greek texts from Alexandria had been translated into Arabic and then into Latin, French and English, a vital stepping stone for science.

Adelard's observations were later followed up by Roger Bacon of Ilchester and Oxford (*c.*1220–92). Bacon laid the foundations for modern scientific thought that we still observe today: hypothesis and objective observations. He rescued the art of chemistry, without which cider making would still be an unfathomable mystery. Alchemy, like cider making, was often conducted behind closed doors. In Paris Bacon was arrested for witchcraft: for experimenting with gunpowder. Alchemy ultimately led to the distillation of cider in order to make cider brandy. Scientific knowledge was a powerful weapon in the right hands.

The seeds of change had been planted in people's minds; playwrights, poets and actors were about to strut their stuff. Orchards and gardens had finally come back together, and apples were the key. For instance, in 1520 the 3rd Duke of Buckingham laid out large pleasure gardens at Thornbury Castle in Gloucestershire. Apples had become ornaments to be tasted with music and poetry, even painted. Wild apples had been tamed, and orchards were statements of wealth and status. Sadly, the duke lost his head for opposing Cardinal Wolsey – Hampton Court politics.

But orchards were very much on people's minds. In 1533 Henry VIII commissioned his chief gardener and fruiterer Richard Harris to take a trip to the Continent. The antiquary William Lambarde (1536–1601) records the planting of the famous mother orchard in his book *A Perambulation of Kent* (1576).

Harris 'fetched out of France a great store of graftes, especially pippins' before which time there were no pippins in England.Teynham, in Kent, was the village where this 'mother orchard' was planted, home territory for Richard Harris, who was born at New Gardens in Conyer, a local smuggling creek. Harris had a distinct liking for cherries.

> He, about the yeere of our Lord Christ, 1533 obtained 105 acres of good ground in Tenham, then called the Brennet, which he divided up into ten parcels and with great care, good choise and no small labour and cost, brought plantes from beyonde the Seas and furnished this ground with them so beautifully, as they not onely stand in most right line, but seeme to be of one sorte, shape, and fashion, as if they had beene drawen thorow one Mould, or wrought by one and the same patterne.

> But as for Ortchards of Aples, and Gardeins of Cheries, and those of the most delicious and exquisite kindes that can be, no part of the Realme (that I know) hath them, either in such quantitie and number, or with such arte and industrie, set and planted.

Today, only one stop down the railway line from Teynham, at Faversham, is the National Fruit Collection which has its own 'mother orchard' at Brogdale. One of the largest collections of fruit trees in the world: over 2,200 varieties of apple, 500 pear, 350 plum, 320 of cherry let alone cobnuts, quinces, blackberries, strawberries, raspberries. A hundred and fifty acres of orchards which, when in blossom, is a remarkable sight, a display that goes on for two months at least. Richard Harris would be pleased.

Dissolution

But Henry VIII had other ideas on his mind apart from apples. By this point Christianity and cider making had been joined at the hip for over 1,000 years, but there was trouble in store for the assiduous monks who had tended their orchards. Henry had them in his sights and over a barrel.

In 1534 the Act of Supremacy made Henry Supreme Head of the Church in England. Henry VIII's marital problems had led to his split from Rome. The wealth and contents of all the religious houses were assessed by Thomas Cromwell's assistants. At the same time a poet and antiquary, John Leland (1503–52), was instructed to visit the monastery libraries to catalogue all their books and manuscripts. These were the Crown Jewels, intellectually, the accumulated knowledge of many centuries.

Then the unthinkable happened; the monasteries were dissolved. Henry privatized religion and monastic buildings became reclamation yards. Some were bought, some were pulled down. Others continued under family ownership. It was a purge, a cultural revolution. Some hard-line abbots were executed to set an example, among them the Abbot of Glastonbury, Richard Whiting, with his treasurer and sacristan. All three were executed on top of Glastonbury Tor in mid-November 1539.

Across the country monks, nuns, friars and canons were sent home, about 12,000 in all. Most were pensioned off. All monastery orchards and cider presses were now privatized, sold off to those favoured at court. Orchards and big houses changed hands. You only have to look at the tithe barns at Glastonbury, Pilton, Bradford-on-Avon or Great Coxwell to understand the scale of these agricultural operations.

Much was lost, especially manuscript books. Worcester Priory had 600 books at dissolution, but only six of them are known to have survived. In York a library of 646 volumes was destroyed, leaving only three known survivors.

The dissolution of the monasteries has been described by John Lawson as 'easily the greatest single disaster in English literary history'. It reminds me of the Chinese cultural revolution. Abbots may have lost their monasteries as well as their orchards, but Thomas Cromwell also lost his head. Far better to stick to orchards and the fruits of one's labours. As a consequence, the skills acquired by monks as alchemists, doctors, cider makers and distillers gradually filtered through into the outside world. The cider brandy cat was out of the bag.

Many families benefited and bought large houses and estates off the Crown. Take Montacute, for example. The only extant part of the Cluniac Priory, founded in 1078, is a tall, elegant gatehouse which is still lived in, and there is a converted cider barn and orchard on the estate.

But of the priory itself, apart from the gatehouse there is no trace. All the fine honey-coloured ashlar stone from Ham Hill was simply recycled and used for building Montacute House. The Phelips family who took over Montacute had interests in glass making and continued the tradition of cider. Even bottling it up.

Herbalists

Having lost so much culture and art in the dissolution of the monasteries, the gentlemen of England saw it as their noble duty to record what their country was all about, particularly as they had split from Rome in such dramatic fashion. Shakespeare did extraordinary things for the stage and the English language, but many lesser mortals also played their part in reshaping England. Not least by simply documenting what they saw in their travels and in their backyards in their own language. Exploration was the order of the day as new horizons opened up both at home and abroad. Herbalists and antiquaries aplenty were fascinated by the natural world, which included apples, pears and cider.

Andrew Boorde (*c.*1490–1549), a Welsh traveller, physician, former Carthusian monk, author and drinks writer visited all of Europe and attended five or six medical colleges. He was also the first man to write books in English on diet and health: *Here foloweth a Compenyous Regiment or Dyetary of health, made in Mountpyller* (1542) and *The Brevyary of Health* (1547), the first medical book in English. Cider has its own remedies: 'Cider is made of the juce of peeres, or of the juce of aples; and other whyle cider is made of both; but the best cider is made of cleane peeres the which be dulcet.'

This is good advice, but a million miles from the modern pear cider which is after all an oxymoron and not to be confused under any circumstances with high-class perry. But not all physicians were impressed with cider.

Boorde continues: 'but the best is not praysed in physicke for cider is cold of operacyon, and is full of ventosytes, wherefore it does ingender evyll humours, and doth savage too much the natural heat of man; and doth not let digestyon, and doth hurt the stomacke; but they the which be used to it . . .'

His final comment is a practical one: 'Cider does little harm at harvest time . . .'

During Queen Mary's bloody reign, many philosophers had to flee for their lives. Then the Elizabethan Age blossomed and it was not just playwrights, poets and actors who benefited. Antiquaries and herbalists were all the rage. Orchards and the natural world beckoned. Men explored science and botany and even tested the hypothesis of God. Ships set sail and brought back extraordinary stories, strange fruits, plants and animals.

All the scientific, philosophical and medicinal knowledge that had accrued in the last 500 years about cider making and distilling was now out in the wider world. Renaissance man was well on his way. This was a sea change which deeply affected cider making. Wine was seen as popish and cider was the new drink to celebrate life and independence.

The only problem cider had was the ongoing battle with ale and beer and the introduction of hops during the reign of Henry VIII. There was fierce debate about hopped and unhopped ale and beer. Hops made the beer bitter but helped keep it in good condition. To some, though, hops were 'the wicked and pernicious weed'.

Herbalists like Dr William Turner (1508–68) were very active and had often learned their trade abroad. Turner, known as the father of English botany, received his BA at Pembroke Hall, Cambridge, then studied medicine in Italy, at Ferrara and Bologna. When he was Dean of Wells Cathedral (1551–3), he established a herbal garden. An advocate of the transmutation of species, Turner was the first to get to grips with botanical classification in Latin and then English. Charles E. Raven in Cambridge wrote that 'Turner, a shrewd observer and an excellent botanist, accepted transmutation as a commonplace event'. Apples, of course, are always transmutating; that's what makes them such fun. Unless, of course, they are grafted or from suckers.

These herbalists and natural philosophers developed a new language with which to express their findings. Turner was also a pioneer ornithologist and wrote the first book on birds. If he could study birds in flight he could certainly study apples.

Not far away at Lytes Cary in Somerset the herbalist Henry Lyte (1529–1607) published *A niewe Herball or Historie of Plantes*, translated from a Dutch work by Rembert Dodoens called the *Cruydeboeck* in 1554. It acquired worldwide renown, in use as a reference book for two centuries, and was the most translated book after the Bible. Rather than plants being arranging in alphabetical order, here they were divided into six groups based on properties and affinities. This was early classification.

Lyte's Cary garden even had forty-four varieties of pear including: 'The Antick Peare, Round Green Peare, Hundred Pound Peare, Capon Peare, Sugar Peare, Ruddick Peare, Russett Sweatter, Orrenge Peare, Warden Peare, Peare Pimpe, Chesil or Peare Nought, Bishop's Censor, Weeten Peare, Great Kentish Peare, Norwich Peare, Winsor Peare, Peare Bell, Peare Laurence, Peare Marwood, Red Gennett, Catherin Peare, Bartholomew Peare, Cary Bridge Peare and Somerton Peare.' Of apples they had 'three score severall sortes'. They must have had exquisite palates and knew the nuances of each variety and how best to store them. Today we have Gala, Conference and Golden Delicious . . . How sad. How boring.

John Gerard (1545–1612) from Nantwich was another botanist. He wrote *Catalogue of Plants* (1596) and *Herball* (1597). For illustrations Gerard used woodcuts from the Continent. More than 1,000 plants and their medicinal qualities were outlined in 167 chapters, in English and Latin. The renaissance of natural history was in full swing.

Gerard's *Herball* again mentions 'lambswool', the wassail drink made from hot, spiced ale or cider with roasted crab apples floating on top, often enriched with sugar, eggs or cream. A bit like a spiced hot posset.

Lambswool and crab apples are also mentioned in *A Midsummer Night's Dream*, II, i. Puck with lips puckered.

If even a crab apple can strut upon the stage of life, what can an entire orchard do? The world of cider was about to strut upon that stage and flex its muscles.

Antiquaries

Antiquaries were also on the crest of a wave, seeking new ways to describe the history of their country. They wanted a voice and made the most of the opportunity. Writing in 1575, William Lambarde, who had recorded the famous mother orchard in *A Perambulation of Kent* (1576), also described the rest of Kent, a county history and topography, the first of its kind. He grew up in Greenwich. Rumour has it that he inherited a manuscript of *Beowulf*. Antiquaries also loved apples, orchards and cider. Then, in 1577, William Harrison (1534–93) wrote *The Description of England* in which he praised the quality of fruit grown: 'In some places of England there is a kind of drink made of apples which they call cider or pomage, but that of pears is called perry, and both are ground and pressed in presses made for the nonce. Certes these two are very common in Sussex, Kent, Worcester.' He obviously did not set foot in Herefordshire, Somerset, Devon, Dorset, Cornwall or Gloucestershire. 'Nonce' means 'for that particular purpose'.

One weighty, leatherbound tome is Camden's *Britannia*, first published in Latin in 1586. William Camden (1551–1623) wanted to restore 'Britaine to Antiquity' and 'Antiquity to Britaine'. He waxes lyrical about Somerset:

'Taunton, it is a neat town, delicately seated, and in short one of the *eyes* of this county . . . the country all about is beautified with green meadows, abounds in delightful orchards, which with the thickness of the villages does wonderfully charm the eyes of the spectators.'

He mentions Orchard Portman and Orchard Wyndham as well as Glastonbury and Avalon. Camden liked his cheese: 'West of Wells, just under Mendippe Hills, lies Cheddar, famous for the excellent and prodigious great cheeses, made there, some of which require more than one man's strength to set upon the table.' Cider and cheddar - a marriage made in heaven.

In the 1695 English edition of Camden's *Britannia* editor Edmund Gibson, mentions that Herefordshire's 'present peculiar eminence is in fruits of all sorts, which give them an opportunity in particular making vast quantities of Syder, as not only to serve their own families (for tis their general drink) but to furnish London and other parts of England. Their Redstreak (from a sort of apple they call it) being extremely valued.'

To get a real flavour of the Continental cider world you have to cross the Channel to the Normandy of 1588. Here the very first treatise on cider was written by a French Protestant, Julian Le Paulmier. Born in 1520 in Agneaux, Normandy, on the Cotentin Peninsula, Le Paulmier was a medical doctor. He became professor and physician to the court of King Henri III in Paris. As a reward for his services, he was ennobled in 1585 by order of the king, acquiring the barony and lands of Vendeuvres and Grentemesnil, on the western border of the Pays d'Auge. He then had time to study wine and cider making. His book *Le Premier Traité du Sidre (The First Treatise on Cider)* was first published in Latin in 1588 under the title *De vino et Pomaceo* and is still regarded as the original treatise on pomology.

This was one of the first books on apple cultivation and storage as well as traditional cider making. Like the authors who followed him, Le Paulmier praised the health-giving qualities of cider. He took cider very seriously, knew his apples and was keen not to mix them. Cider was good for health but not as strong as wine. 'Cider also moistened and corrected the dryness of the bodies solid parts and quenched the thirst.' Le Paulmier also spoke from personal experience:

> For having constantly suffered for three years from heart palpitations and other symptoms familiar to melancholic neuroses, having followed all the known treatments, I got back to my natural good health by going to live in Normandy (to flee the civil wars). I drank nothing but fine cider and within hardly any time at all I was restored to the best of health with no trace left of my former illnesses, which had been pronounced incurable.

The French civil wars were bloody, not least the St Bartholomew's Day massacre of August 1572. Le Paulmier was staunchly Protestant. Thousands were massacred. No wonder he needed cider. His book describes eighty varieties of apple and the various techniques for pressing. Cotentin 'has the best soil for making excellent ciders while those produced in Pays d'Auge are potent and vigorous, but very often dense and badly clarified', a possible reference to apples from the Domfront area, well-known for growing pears.

One of his informants was a man from the Basque provinces, Guillaume d'Ursus. He brought new grafts of apples rich in tannin and acid, ideal for fermentation and, together with Marin Onfrey, another pioneer, worked with Julien Le Paulmier in the area around Cotentin. The French cider connection has often been underestimated in England.

Unspeakable pleasure

Writers played a pivotal role in the Renaissance; and their writings and opinions on orchards and cider were highly valued. The logic of planting orchards was obvious. Orchards provided beauty and profit, both of which made them very attractive to new landowners. Orchards also enhanced their status in the eyes of their neighbours, providing an ornament for their lives and endeavours as well as a fine spectacle to be enjoyed in old age. Both fragrance and fruit.

The Yorkshire Cleric William Lawson (1554-1635) was a dedicated garden writer. In 1618 his book on orchards eventually saw the light of day: this was *A New Orchard and Garden, Or the best way for Planting, Grafting, and to make any pound good for a Rich Orchard; particularly in the North Parts of England*.

Lawson worked on the book for fifty years. When it finally arrived, it was very well received. The second edition included *The Countrie Housewifes Garden*, the first horticultural work written specifically for women. The 'sound, clear, natural wit' of its author was highly praised.

Lawson was well aware of the Classics. He admired 'Plinie, Aristotle, Virgil, Cicero' and many others for wit and judgement. His book not only looked back to the Classics, but forward to the future. An orchard bible. It was as if he had captured the essence and beauty of the

Fig. 5.1 'Perfect forme of a fruit tree', 1618

classical writers and applied it to our own orchards. It gave added impetus, credibility and confidence to those cultivating orchards and gardens. They followed in the footsteps of many esteemed ancestors.

'Let no man having a fit plot plead poverty in this case, for an orchard once planted will maintaine it selfe, and yeeld infinite profit besides', Lawson advises. He also commends the habit of growing 'fruit-trees in hedges, as in *Worcester-shire*'. He sees orchards as equal to cornfields and vineyards: 'And what other things is a vineyard, in those countries where vines doe thrive, than a large Orchard of trees bearing fruit? Or what difference is there in the juice of the Grape, and our Cyder & Perry, but the goodness of the soile & clime where they grow?' Evangelical wisdom, passion and persuasion. These were interesting times and England's real obsession for apples and pears dates from this period.

One clue for this obsession with orchards is in the title of Chapter 16, '*Of Profits*':

> Now pause with your selfe, and view the end of all your labours in an Orchard: *unspeakable pleasure*, and *infinite commodity* . . . and in this Chapter, a word or two of the profit. I count it as if a man should attempt to adde light to the Sunne with a Candle, or number the Starres.

> In *France* and some other Countries, and in *England*, they make great use of Cydar and Perry . . . These drinks are very wholesome, they coole, purge, and prevent hot Agues. But I leave this skill to Physicians.

I love the language: *unspeakable pleasure* and *infinite commodity*. Lawson was in love with his orchards and his handbook on how to establish orchards was mighty useful. The orchard was a metaphor for life. Lawson again:

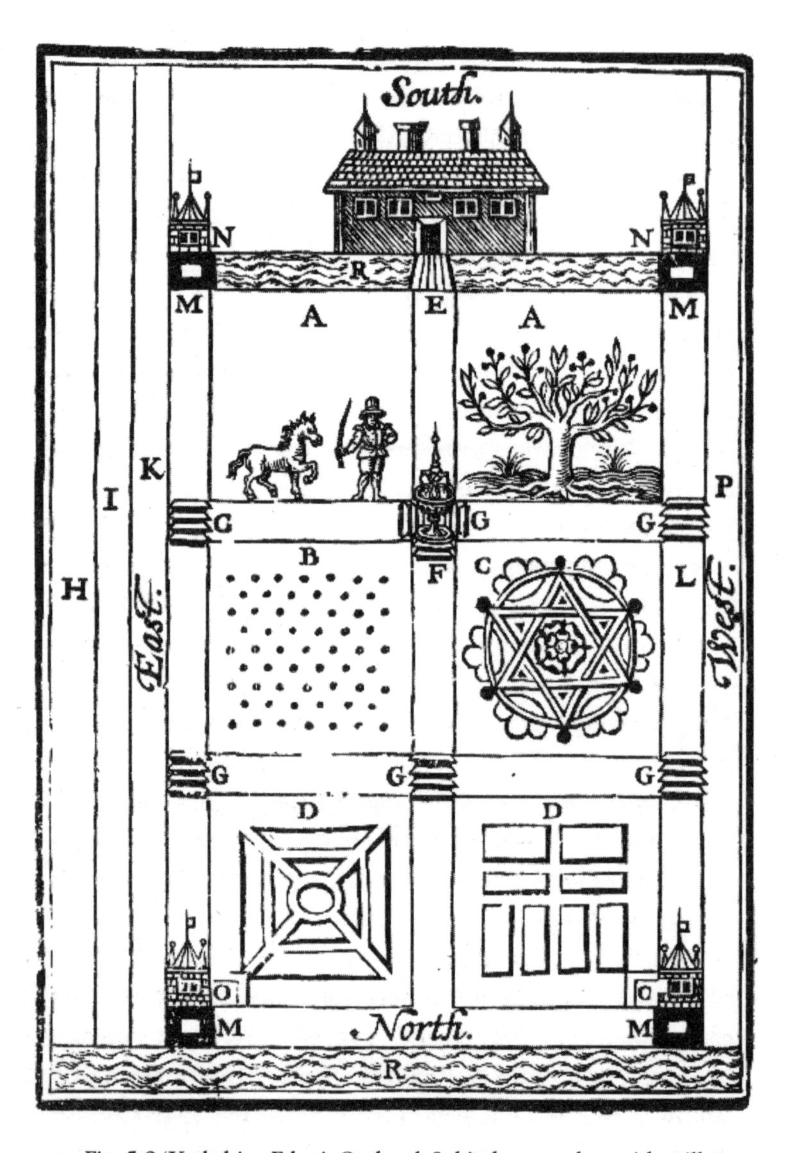

Fig. 5.2 'Yorkshire Eden', Orchard & kitchen garden with still houses and bee hives in all four corners, 1618

> When God had made man after his owne Image, in a perfect state, and would have him to represent himselfe in authority, tranquillity, and pleasure upon the earth, he placed him in *Paradise*. An Orchard is Paradise. What was *Paradise*? but a Garden and Orchard of trees and hearbs, full of pleasure? and nothing there but delights.

What is abundantly clear is that English monks, scholars, doctors, historians, antiquaries, botanists, gardeners, astronomers, alchemists, politicians and plantsmen all travelled on the Continent to satisfy their thirst for knowledge. All brought back samples, cuttings and seeds that could be grown at home.

Another traveller from Suffolk, John Tradescant the Elder (1570-1638), was head gardener at Hatfield House. A fair-sized pile by any standards, Hatfield House was owned by Robert Cecil, Chief Minister to James I, King of England and Scotland. In 1610–11 Tradescant was sent by Cecil to the Low Countries to bring back fruit trees. He was a fine and diligent man described by John Parkinson, the royal apothecary, as 'that worthy, curious and diligent searcher and perseuer of all natures rarities and varieties'.

With his son, John Tradescant the Younger (1608–62), he cultivated apples in Lambeth, then in Surrey. Their 1634 catalogue, *John Tradescant & Son*, lists forty-eight varieties of apple. The Quince Apple, Quince Crab, Dutch Permaine, Yellow Russeting, Pome de Rambure, Dr Barcham's Apple, Pome de Reinette, Grey Costard, Gilliflower Apple, Pidgions Bill, John Apple, Ginnitings, Torne Crab, Blandrille, Summer

Beliboon, Puffing Apple, French Pippin, Sack and Sugar, Black Apple. The list goes on. All dessert apples by the sound of it, their tastes and flavours cherished and espoused.

Tradescant the Elder was a very well travelled. His later journeys took him to Arctic Russia, the Levant and Algiers. He returned to the Low Countries in 1624 and finally went to Paris, but it was in Lambeth that he had his famous cabinet of curiosities, which was very fashionable. Tradescant opened it to the public. Natural and man-made objects. The very first museum in England. To collect is to classify, and to classify is to invest a logical order on what may seem random. Outside in the grounds they had flowers, seeds, fruits and bulbs and inside objects of curiosity.

Mother orchards are similar in that they display nature's DNA. A walled garden with exotic apples and pears at the right time of year is irresistible and – if the gardener isn't looking – you can eat the exhibits. All the sketches of John Tradescant's fruits are now in the Ashmolean Museum in Oxford, along with all his other exhibits in the cabinet of curiosities.

Paradise regained

Born in Yorkshire, John Parkinson (1567–1650) was a herbalist, botanist and founding member of the Worshipful Society of Apothecaries. He was also apothecary to James I and botanist to Charles I. Plum jobs both.

When Charles's queen Henrietta Maria came from France aged only 15, John Parkinson showed her round the private gardens of London. Being somewhat isolated, Henrietta sought solace in flowers and fruit trees. Despite the great difference in age between them Parkinson and the young queen became firm friends. Henrietta Maria, whose mother was a Medici from Florence, was very cultured.

Parkinson had a 2-acre walled botanical garden in Covent Garden with 484 types of plant. He made his own small paradise in Long Acre and encouraged others to do the same. Away from his day job, Parkinson worked on his magnum opus, *Paradisi in Sole Paradisus Terrestris* (1629), the most complete and beautifully presented English treatise on plants of its time, which he dedicated to Queen Henrietta Maria: 'Accept, I beseech your Majestie, this *speaking Garden* that may inform you in all the particulars . . . praying that your highness may enjoy the heavenly paradise . . .'

Speaking gardens and paradise all in the same breath. The illustrated frontispiece says it all: a grand Garden of Eden with tulips, poppies, sunflowers, cactus, crocus, even a pineapple and in the background an orchard laden with fruit, vines, a river with fish, plus Adam and Eve. A wolf skewered on a stake. I delved into my own facsimile copy. Right at the back were chapters on Apples, Pears and Quinces. Parkinson has a light, literary, almost modern style that is very appealing. He hits the nail on the head with his opening lines about apples:

> The sorts of apples are so many and infinite almost as I may say, that I cannot give you the names of all, though I have endeavoured to give you a great many and I thinke it almost impossible for anyone to attain to the full perfection of knowledge therein, not only in regard of the multiplicitie of fashions, colours and tastes, but some are more familiar to one Countrey than to another . . .

As to medicinal uses: 'The juice of Apples likewise, as of pippins and pearmaines, is of very good use in Melancholicke diseases, helping to procure mirth, and to expel heaviness. [. . .] The distilled water of the same Apples is of the like effect.' This is one of the earliest references to distilled cider or *cider brandy*.

What is really fascinating about *Paradisus* is the index. There are three: Latin names, English names and the medicinal *vertues and properties* of each plant or herb are listed. Once an apothecary always an apothecary. For instance: 'to help swoonings; for the Scurvie; to stay the gonorrhaea; to take away vermin and lice in the head. Small pox, strangury and gout.' Always it is the health cures that are as important as the plant itself. It was the perceived medicinal qualities that made apples and cider ever popular. You could even pay your rent in cider – liquid assets indeed.

Parkinson wrote another book, *Theatrum Botanicum* (*The Theatre of Plants*, 1640) with 1,688 pages describing over 3,800 plants, the most complete English treatise on plants of its day. John Parkinson laid the foundations of our obsessions for gardening and orchards. Now it was the turn of cider to reach the palates and dining tables of the highest in the land.

Gerard and Culpepper

The country really was finding its feet. Elizabethan England had flexed its muscles. Goose quills were much in demand for writing and printing presses were working overtime. A new vibrancy was emerging in the shape of lawyers, tradesmen, doctors, entrepreneurs, many of whom had travelled widely and had a thirst for knowledge about the natural world. Their minds were alive and their estates and gardens were becoming increasingly important. The foundations of the middle classes were being laid. Medicine was coming into its own. Tradesmen in towns and cities became very wealthy from wool and weaving woollen cloth, much of which was exported to the Continent. Wine merchants abounded, as did goldsmiths and silversmiths. The liveried trades, guilds and Worshipful Companies blossomed. Expansion was in the air and they wanted new fruit to show off to their friends and to cultivate their wives' interest in gardening. Maybe women did the gardening all along . . . while men put pen to paper.

Another chronicler was Thomas Gerard of Trent (1592–1634). The village of Trent was then in Somerset, but in the county boundary changes of 1896 it was incorporated into Dorset. Trent is the spiritual home of Tom Putt, a large red cider apple which started life further west in Devon at Gittisham, near Honiton. The church in

Gittisham is stuffed full of memorials to the Putt family most of whom seemed to be called Tom . . . One of them, Black Tom (1757–87), was a barrister at Middle Temple and is credited with rearing the original Tom Putt apple. He grew apples and won prizes in Honiton shows. The story goes that Black Tom gave grafts to his nephew, the Rector of Trent, who then propagated it as a dual-purpose apple. Just to confuse things the rector was also called Tom Putt.

In 1633 Gerard produced a humorous and pithy 'Survey of Somerset'. His journey begins in the Forrest of Exmoore: 'a solitarie place, the more commodius for Stagges who keepe possession of it.' He passes Orchard Wyndham, connected to the Wyndham family. Then Gerard passes Wellington, where he waxes lyrical about the soil:

> by the Riverside have ye a most pleasant and delectable walke to Taunton . . . for beside the River which is here very cleare, you have the prospect of many faire orchards, gardens and cherry gardens. Of which there are a great number hereabout, and there cherries ripe welneere as soon as at London, for the greet is rich redd earth which produceth all fruits not only in great plentye but very early.

And when he gets to Taunton, 'A fair and pleasant towne this is, I assure you', but on the south side of the town, he takes a liking to Orchard Portman, and there are fantastic etchings of a painting of the estate. Apple trees were everywhere. The estate has now been swallowed up by Taunton racecourse.

'The whole Countrey thereabouts seems to be orchards, in so much that all the hedgerows and pasture groundes are full fraught with fruit trees of all sorts fittinge to eat and maker cider of.'

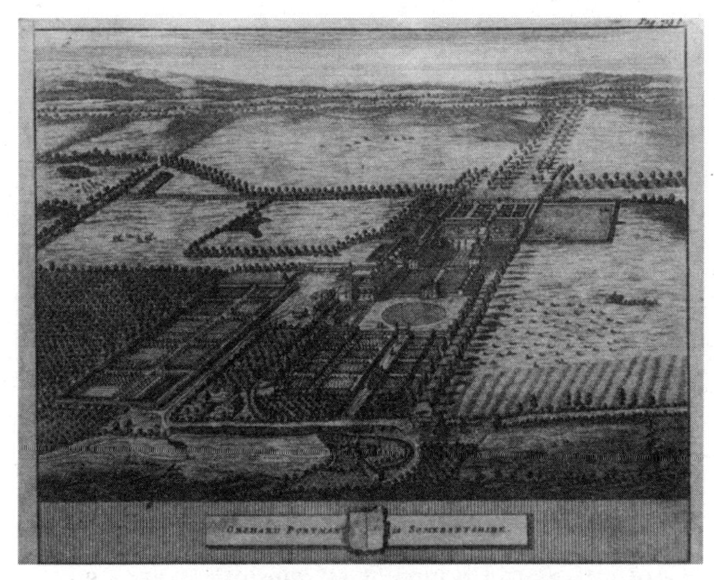

Fig. 5.3 'Orchard Portman and its extensive orchards', Taunton, Somerset, 1707

As always, antiquaries had a good eye for the lie of the land. Farming fortunes depended on it. No wonder the Vale of Taunton was so plentiful and Taunton cider flourished.

Nicholas Culpeper (1616–54), a Cambridge-educated botanist and herbalist, was apprenticed to an apothecary before practising as a doctor in London's Spitalfields. During the Civil War, he was accused of witchcraft but later fought on the Parliamentarian side at the first Battle of Newbury where he carried out battlefield surgery. He was himself wounded in the chest by a bullet. He went on to write *The English Physitian* (1652), the *Complete Herbal* (1653) and a detailed book on medical astrology, *Astrological Judgement of Diseases from the Decumbiture of the Sick* (1655).

The basis of his works was reason and observation rather than tradition, combining planets with Galen's philosophy of humours. Culpeper was a widely read source for medical treatment. He used mild cider vinegar as part of a herbal concoction for a hair tonic.

Culpeper recommends apples for 'hot and bilious stomachs, inflamations of the breast and lungs and asthma'. Also a cold poultice of boiled apples mixed with milk for gunpowder burns. Very useful during the Civil War.

In his *Complete Herbal*, Culpeper records a recipe for wassail cup: 2–3 cinnamon sticks, 3 blades of mace, 4 cloves, 1 teaspoon of nutmeg, 1 ginger root, 4 apples, 4 oz. of sugar, ½ pint of brown ale, ½ pint of cider.

Cider wassail was an almost religious affair with a wooden altar and special cups, as if it was a pagan communion. Some also had spice boxes set on top of the altar. I have seen one Elizabethan wassail bowl made from lignum vitae, found round the back of an old cider farm. It was about 12 inches high, of very dark wood and slightly at an angle. It had the genuine patina of age and had been turned on a lathe. It held about 2 quarts.

Wassail was a good excuse for a communal party whether in the large house, or in a village or a remote farm. John Aubrey (1626–97) notes that on Twelfth Night, men 'go with their wassail bowl into the orchard and go about the trees to bless them and put a piece of toast upon the roots, in order to encourage it'. Some farmers even took the wassail bowl into the stable and offered it to the oldest cow. Wassail brought people together on Twelfth Night with wassail maids offering cider. Cider was deeply embedded in ritual. The Catholic Church had gone underground but the need for ritual was keenly felt, particularly when food and survival were uppermost in people's minds. God still existed and so did their reliance on good weather and a bountiful harvest. Winter solstice, Christmas, and New Year wassail all had their place in people's hearts and homes.

For others, particularly in the South West, it was the high seas that caught their attention, the discoveries of land and riches far away both east and west. The age of exploration and colonization had just begun. And, like all good mariners, they needed good cider.

6

MARITIME CIDER

Elizabethans were great explorers and navigators. They liked cider, perry, syllabub, posset, verjuice, madrigals, lutes and wassail cups. As well as exercising their newfound freedom with language and maps, charting the unexplored world and crossing boundaries. Elizabethan English was a language of new ideas, rich veins of tragedy, humour and wit. History was being written, rewritten and invented. Expansion and improvement were new philosophies. The nation was on the move.

Orchards were places of education and refinement, poetry, intrigue, romance and secret trysts. How many plots in Shakespeare's plays depend upon conversations overheard in gardens and orchards? Shakespeare talks of codlings, pippins, leather coats, apple-johns, crabs and *bitter-sweetings*, which were cider apples.

In *Romeo and Juliet* (ii. 4) Mercutio says:

> Thy wit is a very bitter-sweeting; it is a most sharp sauce.

Sauce indeed . . . good for the goose, good for the gander. There is also a fourteenth-century *bitter swete* in the Middle English poem *Confessio Amantis* – 'The Lover's Confession' – by John Gower (1330–1408):

For all such time of love is lore
And like unto the bitter swete,
For though it thinke a man first swete,
He shall well felen atte laste,
That it is sower, and maie not laste.

Crab apples were familiar to Caliban in *The Tempest* . . .

Let me bring thee where the crabs grow . . .

And in *Love's Labour's Lost*:

When roasted crabs hiss in the bowl.

Another Elizabethan habit, particularly prevalent in Devon (where seafaring and cider were never far apart), was swashbuckling. Pirates and privateers who set sail and explored the unknown world.

Cider was a vital part of victualling. Piratical cider. Maritime cider. Swashbuckling cider. But voyages of exploration came at great cost. In 1595 Admiral Sir Richard Hawkins (1562–1622) declared that scurvy was 'the plague of the sea' and that 'in 20 years at sea he knew of 10,000 men who had died of it'. Scurvy is a very unpleasant condition whose symptoms include suppurating sores, open wounds, swollen joints and cracked lips. Hawkins, who sailed against the Armada and discovered the Falkland Islands, urged 'learned men' to write about it.

His father, Sir John Hawkins (1532–95), also fought against the Armada and went on treasure-hunting escapades with Sir Francis Drake. Good high-tannin cider in solid casks kept far better than beer in the tropics and helped to prevent (or at least delay) the onset of scurvy. On a voyage in 1564 John Hawkins took 40 tuns of beer, 35 tuns of cider and 40 butts of Malmsey at £6 a butt. A tun is four hogsheads (about 208 gallons), and Malmsey a sweet Madeira wine.

Sir Francis Drake (1540–96), another fearsome Devon sea dog, was not averse to piracy and exploration. Born on a farm with cider orchards called Crowndale, a mile south of Tavistock, he knew all about cider. In 1581, having circumnavigated the world, Drake was knighted on the *Golden Hinde* (not, as is widely reported, by Queen Elizabeth but by a French diplomat). Drake then bought Buckland Abbey in Devon which was famous for its cider.

The fact that so much cider was taken to sea in Elizabethan times is highlighted by John Parkinson in 1629, who draws a clear distinction between dessert apples and cider fruit:

> Some sorts are best to make Cider of, as in the West Countrey of England great quantities, yea many Hogsheads and Tunnes full are made, especially to bee carried to Sea in long voyages. And is found by experience to be of

> excellent use, to mixe with water for a beverage. It is usually seene that those fruits that are neither fit to eate raw, roasted, not baked are fittest for cider to make the best . . .

In 1790 William Marshall records that 'One of the orchards at Buckland Priory is said to be the oldest in the country, and this is spoken of as being about two hundred years old. Nevertheless, this Orchard is still fully stocked, and in full bearing!' Today the tithe barn, which is 95 yards long, lies empty apart from a large cider press and a few barrels.

Drake also liked sherry. Dry sack to be precise. In 1587, when 'singeing the King of Spain's beard' in Cadiz, he 'liberated' 2,900 butts of sherry at the docks destined for South America (313,200 gallons, or 300 tons). Drake's escapades delayed the Armada by a year. One cider maker I knew in the Tamar Valley always advocated putting a bottle of sherry into a gallon of cider to make the party go with a swing. Drake would have liked that.

At one point off Panama, Drake joined forces with a French Protestant privateer and explorer called Guillaume Le Testu (*c.*1510–73) who had several casks of wine and Normandy cider on board. They desperately needed water. They had a drink, then both went off to ambush a Spanish silver mule train. Two hundred mules and 30 tons of gold and silver were captured, half of which was buried. Guillaume Le Testu was wounded and captured by the Spanish who cut off his head. Crime does not always pay.

Cider and scurvy

Elizabethans certainly knew how to enjoy life. Currying favour at court required not only tight hose but a seaworthy ship, bold, daring plans and a deep purse. And perils there were many. As Phocylides of Miletus said much earlier: 'Shew mercy to those that are shipwrecked, for the sea, like fortune, is a fair but *fickle mistress*.'

Away from court and on the high seas with letters of marque, sea dogs roamed the great oceans in small vessels, seeking out new lands for their queen and a fortune to be made with booty. Yet reality was harsh. Life on the high seas was often short and cruel. On long voyages no amount of poetry and love sonnets could save men from dying. But scurvy was not just a British problem. In 1499 Vasco da Gama, on his voyage to India, lost 116 of his crew of 170 and in 1520, around Cape Horn, Magellan lost 208 out of 230. All mainly to scurvy.

In England cider was perceived as a very practical solution, at least in the short term. Long voyages to the tropics were more problematic. Many writers proclaimed the good services of cider at sea and the occasional, very successful use of lemons and oranges. Tannin helped preserve cider from degradation which in turn preserved the crew's morale on long voyages. Expedition cider had to be very good to survive the tropics. It was often observed that while the cider remained fresh, scurvy did not develop. Sir Hugh Plat (1552–1608), known for *The Jewell House of Art and Nature*, had a leaflet in 1607 entitled 'Certain Philosophical preparations concerning food and beverages for seamen in their long voyages'. He furnished Sir Francis Drake and Sir John Hawkins on their last voyage. As an alternative to meat and fresh victuals he advocated macaroni from Italy and couscous from Barbary. As for beverages, he advocated 'that all barrels be cleansed with "philosophical fire" being of a sympathetic nature. By this means also Wine, Perry, Cider, Beer and vinegar may be safely kept at sea, for any long voyage, without fear of growing dead, sour or musty.'

Philosophical meant 'scientific'. Natural philosophy was a wide subject. The best way to cleanse barrels was by using sulphur (brimstone) with spices to fumigate the barrels. Basic science.

In his visionary book *New Atlantis*, the philosopher Francis Bacon (1561–1626) recommends that travellers be greeted with cider and oranges. Cider and perry are notable beverages during voyages, 'an assured remedy for sicknesse taken at sea'. In an earlier 1605 work Bacon tells us that cider makers 'take great care not to bruise or squeeze the apples till they have lain together for a while in heaps, and so ripened by mutual contact'. Thus making a rich, deep, strong cider. Much better than most cider made today.

One man who also knew about scurvy was Sir James Lancaster (1554–1618), a sea dog from Basingstoke in Hampshire. He sailed for the East Indies from Plymouth in 1591 and returned in 1594, having lost many men to scurvy. In 1601 Lancaster was employed by the newly formed East India Company. His command, the *Red Dragon*, was a 38-gun ship with a crew of 200 men. The daily ration per man was three small loaves of bread a day, cheese, butter, fish (dried salt cod) or one pound of meat a day (salted or pickled beef or pork), half a pint of peas or beans, a pint of wine, a quart of beer or cider and aquavit. Fresh vegetables were hard to find. That was the problem. The sheer quantity of cider taken on board must have been phenomenal. The ships took on rations for twenty months. One pint of cider per man per day: that's 365 pints a year. With 200 men on board *Red Dragon* that is 9,125 gallons, or 40 tons of cider a year. So for twenty months that was 66 tons of cider, not counting the weight of the casks themselves. And that was only one vessel. Lancaster observed that men aboard *Red Dragon* were often healthier than those on other ships.

James Lancaster's fleet of four ships departed Torbay in April 1601. They were delayed by unfavourable winds, followed by a month in the doldrums. Never a good thing. They arrived in Table Bay, on the southern tip of Africa, in November. The crew of *Red Dragon* were in good health but the other men on smaller ships were badly affected by scurvy. A fifth of them died. The only difference was that Lancaster insisted his men take three spoonfuls of bottled lemon juice every day, a trick he had learned from the Dutch who in turn had learned it from the Portuguese. Oranges and lemons were scarce in England but available from Portugal. This was the first large-scale experiment with lemon juice. But the beneficial effects of the cider were also noted. Cider may well have worked for three months before losing its efficacy in the tropics.

Lancaster's fleet progressed to Madagascar and stocked up on lemons. Then it headed for the Nicobar Islands, Sumatra and Bantam in Java. Pepper, spices and calico were brought back home. They set up a factory in Bantam. They were away for three years and, upon his return, James Lancaster was knighted.

Richard Hawkins also wrote about oranges and lemons on voyages back in 1593: 'This is a wonderful secret of the power and wisdome of God, that hath hidden such an unknown and great vertue in this fruite, to be a certaine remedie for this infirmitie.'

But cider was still very effective. In 1676 John Worlidge, a Hampshire man who knew Portsmouth well, commended cider. 'For its specifick Vertues there is not any drink more effectual against the Scurvie.'

In reality cider appeared to be very useful for preventing scurvy on voyages in the northern latitudes but less so on long voyages of over a year in the tropics. Cider did, however, cure many a sailor coming ashore who had scurvy symptoms once they reached the home port. In other words, go down your local pub and drink as much good high-tannin cider as you can. Hence cider's high reputation. It depended very much on the quality of cider and the way it was kept. The ambient temperature would be important. Cider keeps better at low temperatures, so for cod fishing and Newfoundland voyages, cider was fine, but for cruising in the tropics less so. Cider was then only a partial cure and preventative.

But why did cider work? These days we take vitamin C for granted. It is a necessary nutrient that helps the body absorb iron and produce collagen. If the body does not produce enough collagen, tissues will start to break down. Symptoms of vitamin C deficiency can start to appear after twelve weeks. Early signs include loss of appetite, weight loss, fatigue, irritability. Lethargy sets in. Not a good state of affairs.

The naval expedition that really highlighted the scurvy problem was George Anson's voyage around the world of 1740–44 with six vessels. Of the original crew of 1,854, about 1,300 men died a slow, painful death from scurvy, a catastrophic and unnecessary loss.

One of my Irish ancestors called James Crohane was on board one of these ships, the HMS *Severn* as captain's secretary. He survived, but only just. *Severn* never rounded Cape Horn. The men were so devastated by scurvy that they couldn't handle the sails in a storm. Very reluctantly they turned back and limped into Rio de Janeiro. Of 300 men, 158 had died and of the remainder 114 were too sick to help, leaving just 30 men and boys to work the ship. This does not count the invalids and marines, nearly all of whom died. The ship's log makes terrifying reading.

Anson was declared a hero and became Admiral of the Fleet, but thousands died along the way. There had to be a solution to scurvy. It was a crisis that involved the navy from top to bottom.

The cure

Eventually things did start to happen. There was national outcry and even the navy realized it had to shift its backside, ordering two young naval doctors to apply their minds to the problem. Both experimented with cures just as James Lancaster had done in 1601. It was now 1746 . . .

Of course, no one knew about the existence of vitamins in those days, but they did know all about the need for good fresh victuals (pronounced vittles), i.e. food and vegetables. Rock Samphire and scurvy grass or spoonwort were also used as a cure. The Revd George Moore recorded the purchase of 'a pint of scurvey-grasse' for 1*s*. in 1662.

But a foolproof cure eluded them so naval doctors experimented with various concoctions including cider, observing the progress of their patients. As with all good science, accurate notes and an open mind were essential. In the end they saved many lives.

In 1746 Edward Ives (1719–86), a surgeon from Lymington, Hampshire, was serving on HMS *Yarmouth*. Having lost many men to scurvy on a previous voyage in the Mediterranean, Ives prevailed upon the admiral to let him have some cider. The admiral condescended. Anything was worth a shot. He took on board several hogsheads of best South Hams cider: a quart per patient. Of 500 men not a single one was lost while the cider lasted. The cider must have been fresh to start with. At least one would hope so. And that was why it worked.

Ives's precise words make interesting reading. 'I had myself formerly experienced the good effects of apples, it was reasonable to assume that cyder must certainly be of service. This suggestion agreed with some accounts the Admiral had heard from others; and he with great readiness, bought and put under my care, several hogsheads of the best South Hams Cider. During the next cruise each scorbutic patient had daily a quart or three pints of cyder. They also had to drink 3 quarters of a pint of seawater in the morning twice a week. On the channel cruise they had as many patients as other ships, but whereas the other ships buried at sea 20, 30, 40, 50 men and upward, their ship *Yarmouth* only buried two or three and that was at the end of the voyage.' Cider had run out ten days earlier. Many men were hospitalized but there was a marked improvement.

Something was obviously happening. If they left port in January the cider would have been from the year before, i.e. autumn 1745. That meant it was getting on for eighteen months old by the end of the voyage. The cider had been through two summers, which would have led to a decrease in vitamin C when the weather warmed up.

So any beneficial effects of cider tailed off, which is what Ives observed. It would have been better if they had treated all the crew with cider every day. But with some 500 men on board that would have been expensive and taken up more room in the hold. Regardless, it was a step in the right direction; cider had saved many lives. Mr Ives wrote a letter to Mr Lind conveying his observations. In 1753–7 Ives was in the West Indies on *Kent* and used rum and lemon juice to good effect.

Local apples were still used on board ships. Hampshire

had the Hambledon Deux Ans, which keeps for two years. Just the sort of apple you need on long voyages. Hunt House apples, from near Whitby, were favoured by Captain Cook on long voyages, while Ribston Pippins have six times more vitamin C than Golden Delicious.

Fast-forward to 1978 and the cider historian Roger French, with the help of Dr Paul Luzio, analysed the vitamin C content of modern ciders. Roger French reckoned that the eighteenth-century vitamin C content of cider was as high as 40–50 mg per litre. They looked at Strawberry Norman, which gave a cider of 33.8 mg of vitamin C per litre. Scrumpy Jack from Symonds, a traditional 7.5 per cent cider from Stoke Lacy, gave a reading of 47 mg per litre. So all you needed for your seafaring daily ration of vitamin C was one litre a day of old-style Scrumpy Jack. Paid for by the navy!

But to return to the scurvy saga of 1747, James Lind (1716–94), a Scottish surgeon on board *Salisbury*, also conducted a trial experiment with twelve men suffering from the condition. Two got lemons and oranges, two more a quart of cider and the rest received popular anti-scurvy medicines such as Purser's pills, Vinegar, 'elixir of vitriol' (a mixture of sulphuric acid and alcohol), and a patent medicine containing antimony and balsam called Ward's Drop and Pill (known for its laxative effects). A month later, when the ship sailed into Plymouth, those on fresh fruit were cured. Those on cider had improved. The others not at all. In conclusion, Lind states that next to oranges he thought cider had the best effect. But he agreed that the casks of cider he possessed were not very sound, liable to become sour or acetic. In other words, it had turned into vinegar and had lost much of its vitamin C. If he had had fresh, strong South Hams cider the results might have been much better. Lemons were easier to store if you could get hold of them, but that wasn't always easy. Hence our links with Portugal.

In 1753 Lind wrote *A Treatise on Scurvy*, establishing the importance of fresh fruit and vegetables. However, the Admiralty did not implement his recommendations and scurvy remained a significant cause of sickness in the fleet, particularly when blockading France for months on end. The navy had not listened to their own medical experts. A despatch of 25 September 1759 from Admiral Hawke's squadron shows that when a resupply ship arrived with a shipload of apples, law and order broke down completely. Apples were 'immediately plundered by the men before the pursers could take any account of it'. It shows the innate power of apples over naval discipline.

It eventually took a Scottish doctor, Sir Gilbert Blane, Physician to the Fleet (1779–83), to persuade the Admiralty to change its ways. A rare achievement. In 1785 he wrote a book *Observations on the Diseases of Seamen*, but it was not until 1795 that the navy took on board his message loud and clear. A staggering forty-three years after Lind wrote his paper, and one year after Lind's death, the Admiralty issued an order for lemon juice to be issued to sailors.

The problem was that no one knew the inner workings of lemons, limes, oranges or apples. How could they? Vitamin C – ascorbic acid – was only identified in 1912 as the key ingredient, *vitamin* or vital mineral, which cures and prevents scurvy. So for hundreds of years, doctors, ship's surgeons and old sea dogs were going in the right direction but working in the dark. The Admiralty was very sluggish, particularly as the West Indies had plenty of limes. *Triste tropiques*.

But the real invisible problem with vitamin C is that the body cannot retain it, so it has to be constantly topped up, which is where cider comes in. Fruit is at best seasonal. Modern industrial cider has very little if any vitamin C left in it. Making apple concentrate and pasteurising destroys it, just as in the tropics.

So if you ever sail from Plymouth, take high-quality South Hams cider with you plus as many lemons and oranges as you can find. And if you return with scurvy, simply visit the local cider houses closest to the dockyard. But if you were really ill you went 'Up the Creek' to one of the Naval hospitals. Old Naval slang for being very ill and lucky to return alive. Others went to hulks . . .

Sea dogs

Many sea dogs had estates and orchards in Devon. They vied with each other as swashbucklers. 'Liberating' salt cod helped their finances, because cod was a vital part of the European diet and a key economic driver in the exploration of Canadian and American coastlines.

Like high-tannin cider, salt cod is an acquired taste. The dried cod, which is as hard as planks, is soaked before use.

Softening up is therefore highly recommended. As that wise wit and raconteur Benjamin Franklin (1706–90) observed while giving advice to those undertaking a sea voyage: 'You will find that cyder is the best liquor to quench the thirst generally caused by salt meat or salt fish.'

Newfoundland. Rumours were that Thomas Croft (1442–88) of Herefordshire, a customs officer for Bristol, visited secretly in the 1470s. John Cabot later 'discovered' Canada in 1497. By 1500 Newfoundland cod was landed in Bristol and within ten years many ships sailed there from West Country ports including Dartmouth and Topsham. According to one account 'the sea was full of fish, taken not only with nets but with fishing baskets'.

Taking cider to sea was common practice. In 1520 a vessel from Bayonne, with trade links with Helston, in Cornwall, took 40 butts (or 80 hogsheads) of 'the best cider that can be found' to the Newfoundland fisheries. The owners of the vessel, Michael de Segure and Matthew de Biran, were cider makers and orchardmen. Cider was obviously in demand on the high seas.

By 1540, the English were sailing to Newfoundland in much larger numbers, from Bristol, Bideford and Barnstaple. Cider was vital for salt cod fishermen and salt cod was vital for the navy. One document in the British Museum estimates that in 1545 the *Mary Rose*, with a crew of 700, required 18,000 fish per week for eight weeks. That is a lot of salt cod. Eighteen per cent protein. By 1578, England and Portugal each had fifty ships engaged in the fishery, and France and Spain 150 apiece. Things were hotting up.

Catholics ate fish on Fridays so there was a guaranteed market in Portugal and Italy. Cider and salt out, salt cod to Lisbon then port wine, olive oil and dried fruit back. Very profitable.

In 1583 Sir Humphrey Gilbert (1539–83) and Sir Walter Raleigh (1552–1618) were engaged in an expedition to Newfoundland with a crew of misfits, criminals and pirates. They sailed with a fleet of five vessels to Newfoundland armed with letters patent, all set to claim NEW- FOUNDE-LANDE for Queen Elizabeth. On arriving at St John's in August 1583 they were met with a hostile party of several different nationalities. But Gilbert waved his letters patent about and, in a formal ceremony, took possession of Newfoundland for the English Crown. Some turf was cut to symbolize the transfer of possession and he was then presented with a dog which he named Stella after the North Star. Queen Elizabeth now had a new overseas possession and loads of salt cod. Sadly, Sir Humphrey drowned on his way home. He was said to be reading a book at the time.

Upon his return, Sir Walter Raleigh described the Newfoundland fishery as 'the mainstay and support of the West'. He should know. Sir Walter grew up in Hayes Barton, near East Budleigh, in East Devon. He also had Sherborne Castle under his wing. His Dorset estates, according to Gerard, were 'beautified with orchards, gardens and groves of much delight'. Sir Walter also had estates in Ireland and introduced cherries to Waterford from the Canary Islands. Plus potatoes and tobacco.

Fig. 6.1 'Barent Langenes map of Terra Nova', Newfoundland, 1602

Two years later, in 1585, another sea dog sailed to Newfoundland: Sir Bernard Drake (1537–86) of Ashe near Axminster. Sir Bernard was less well known than Sir Francis, but the same piratical streak was definitely there. Sir Bernard captured several Spanish vessels and 60,000 quintals of dried fish (a phenomenal amount: about 2,650 tons of salt cod) as well as plundering other vessels carrying gold and sugar. The following year the Spanish decided to steer clear of Newfoundland. The Portuguese fishery never recovered. Two years later the Spanish sent the Armada up the English Channel. Cod wars. The gloves were off.

Cod fishing was just like a goldrush. Around 1,600 English fishing captains still reported cod shoals 'so thick by the shore that we hardly have been able to row a boat through them'. Hundreds of vessels went to Newfoundland every summer, among them those also from Portugal, Spain, the Basque region, France and the Channel Islands. Rich pickings for pirates. The fishermen also set up temporary summer camps on the island. Devon cider for cod fishing was a major part of the rural economy that fuelled long sea voyages and kept sailors in good spirits when ashore.

Anything between 250 and 400 English ships were involved with the fishery at its height, along with 6,000 to 10,000 men. A lot of cider was needed. Dartmouth provided 20 per cent of these men and ships. The fishery was so important that these men were exempt from being press-ganged into the navy in time of war.

In 1618 Sir Richard Whitbourne (1561–1635) set out from Exmouth with yet another expedition to Newfoundland armed with SIDER. He had fought against the Armada and had sailed to Newfoundland many times. A practical sea captain, he became de facto Governor of Newfoundland. When fitting out his ships the list of supplies to be loaded makes fascinating reading. Whitbourne took 26 tunnes of beer and cider at 53*s*. 4*d*. the tunne which comes to £69. 6*s*. 8*d*. – 16 per cent of the whole bill. He also had to supply cider for his men in Newfoundland. He took two hogsheads of very good English beefe (£10) and two of Irish beefe and, for £5, 2 cwt of cheese. Lucky mice. Countless other useful objects and foodstuffs and there, at the tail end, next to 2,000 orlop nails, is a hogshead of aqua vitae. No bibles. No wine. Cider instead.

Pirates and pilgrims

Exploration was becoming big business. The East India Company was founded in 1600. Each seaport made a small fortune fitting out expeditions and then reaped the rewards. Gunpowder, cider, salt and provisions on the way out. Salt cod, timber and plundered booty on the way back. Settlements were then established in America and Canada.

By 1620, 300 fishing boats worked the Grand Banks off Newfoundland, employing some 10,000 sailors. What they really needed was not just cider but apple trees, to make a permanent settlement with orchards. But the conditions – sea fog and wind – didn't make it easy.

In 1627 the *Ark of Avalon* sailed from Dartmouth. Many of the half-timbered houses in Dartmouth date from this time, funded by cod and cider. By the 1670s there were 1,700 permanent residents in Newfoundland and another 4,500 in the summer months. In 1675, seventy percent per cent of the boats on the Grand Banks were from Devon. The uncharted quantities of cider drunk on the high seas or in bothies, shabeens and pubs in Newfoundland must have been prodigious.

A survey by Dean Milles in 1750 estimated that Devon produced 170,000 hogsheads of cider a year, the equivalent of 10 million gallons; the population was around 300,000. Devon was indeed a top-notch cider county, with some parishes making large quantities. West Ogwell's cider was 'remarkably good', Gittisham's was a 'good masculine cider' and Ottery St Mary's was 'a good cider improving annually'.

Cider was also going up the country in the opposite

direction. In 1720 Daniel Defoe states that in East Devon 20,000 hogsheads are sent every year to London, an enormous amount of cider. And that is just East Devon, let alone South Hams cider from Dartmouth, Kingsbridge or Salcombe. Defoe says that most of the cider sent to London 'is bought there by the merchants who mix it with their wines, which if true, Is not much (credit) to the London vintners. But that is bye the bye.'

Also there were plenty of cider houses in London – midnight cider palaces and opera houses – J. M. W. Turner lived opposite one. But on the other side of the Atlantic something else was going on.

Back in 1620 at the same time as Richard Whitbourne left New-found-land, an odd bunch of earnest religious dissenters, known as the 'Pilgrim Fathers', set sail from Plymouth on board *Mayflower*. Bound for America, they clutched their bibles and each other during the voyage across the pond. There was a rumour that one hogshead of cider was smuggled on board. They also took a long iron screw thread which could be used to make a rudimentary cider press. In a severe storm the screw thread was used to brace the ship internally when its timbers were creaking and she was about to break up and sink. Cider engineering to the rescue.

After sixty-five days land was sighted. By December, most of the pilgrims and crew were suffering from scurvy. Out of 102 pilgrims, 45 did not survive that first winter. They may well have craved apples but apart from a few indigenous wild crab apples there were no cider apples in America and certainly no cider. This obviously had to be remedied . . .

Colonial cider

Maritime cider then took a leap of faith and expanded its philosophy south of Newfoundland into very different terrain. From salt cod to Cape Cod. The climate of Virginia and New England was much warmer and more conducive to apple growing. But the Pilgrim Fathers, for all their religious intensity, were not the first settlers. Not by a long chalk. Sir Walter Raleigh had encouraged settlements in North Carolina in 1585 and 1587 but without success. One colony disappeared entirely.

Twenty years later, in 1607, settlers tried again, this time at Jamestown in Virginia. But not everything was rosy. After the first winter only 38 out of 108 settlers were still alive. Next winter the population dropped from 500 to 60 in six months. Scurvy had reared its ugly head as well as starvation and disease. The settlers desperately needed apples and they needed cider to tide them over the winter.

One well-documented case of apple trees being sent out from England is from 1620, on *The Supply*, a companion ship to *The Mayflower*. She left from Bristol and landed in Berkeley, Virginia, on 29 January 1621. Berkeley was a settlement founded in 1619 with thirty-eight settlers and named after Richard Berkeley. On board were fifty new settlers, including Richard Smyth, his wife and two sons.

One letter reads:

> Scions were sent to Jamestown on a ship called *The Supply.* Mr Smyth hath this day drawne out of his nurseryes a great number of yonge stocke and of apples tress grafted

with pippins, pearmaynes and other the best apples. Where he hopeth for his owne humoir and affecon sake therin you have somewhat the more care of, as also of the bagg of abricote, damosell, and other plumstones he nowe sendeth . . .

Your assured loving friendes,
Stoke Sat 10th September 1620
Richard Berkeley and John Smyth.

Berkeley came from Stoke Gifford, north of Bristol, and Smyth was a Gloucestershire antiquary from North Nibley, near Dursley. Nibley House was 'adorned with gardens and groves and a large park well wooded'. Both men were shareholders in the Berkeley Hundred Colony in 1618.

Apple trees for the New World were therefore very important and once established grew well. Two years later, in 1622, in compliance with a request from the colonists, the Virginia Company despatched a pinnace with wheat and barley, garden seeds and scions of fruit trees. The first shipment of honey bees to America for the pollination of apples is also recorded in Virginia the same year. By 1629 Captain John Smith says some settlers have planted orchards 'that have prospered exceedingly'. Not just apples but 'pears, peaches, apricots, vines, figs and other fruit'.

Another early reference to apples is from *Marmaduke*, setting sail on 16 September 1623 from London bound for Jamestown. On board were apples paid for by Mr Maddocks. There were two shipments: one valued at £8.

6*s*. 6*d*. (a lot of money) and a second at 16*s*. 6*d*. There was also a payment of £3. 4*s*. to the aqua vitae man and an order for one hogshead, fower (4) barrels and 3 roundlets to George Harryson of James Citty.

The *Marmaduke* was well known. Two years earlier, in 1621, she had carried fifty-seven unmarried women of good character, shipped out by the Virginia Company, who paid for their transport and provided each of them with a bundle of clothing and other goods. Any colonist who married one of the women was responsible for repaying the Virginia Company for his wife's transport and provisions. All the women were snapped up and married on the spot. Looking after orchards, cider and distilling would have been on the cards. Settlements blossomed.

In 1626 Richard Kingsmill, of Archer's Hope, James City, planted a pear orchard and made from fruit gathered there some forty or fifty butts of perry. In addition to his house at Jamestown, George Menefie maintained a plantation called 'Littletown' where he had orchards of apple, pear, cherry and peach trees, and a garden noted for its rosemary, thyme and marjoram. Captain Brocas of the Council kept an excellent vineyard on his plantation, in Warwick County, patented in 1638. Richard Bennett, of Nansemond River, developed an apple orchard. In 1648, he had made twenty butts of cider. Cider making was now endemic.

Further up the east coast in 1629 in Massachusetts, William Blackstone planted apple trees in Boston. A dissenter, he was now a minister to the Plymouth settlers.

Religion was deeply embedded in people's minds and this caused a few problems. In America religion and apples were never far apart. In 1630 the Endicott Pear Tree was planted. Located in Danvers, Essex County, Massachusetts, it is said to be the oldest living cultivated fruit tree in North America. As a sapling it was brought out on *Arabella*, flagship of the Winthrop Fleet. Revd John Endicott, a zealous Puritan who planted the pear tree, came from Chagford in Devon. He also established the first known nursery for apples and pears in Salem Village. Remarkably the Endicott Tree is still alive, just.

Cider and orchard philosophy were alive and kicking. The orchard cult had been succesfully transplanted from England. Cider flowed. Each autumn America would have smelled just like the West Country.

Presidential cider

American cider was not just religious, it was also deeply political. When George Washington ran for the Virginia House of Burgesses in 1758, his campaign expenses included eight quarts of 'Cider Royal'. He won the election and ended up serving more than 144 gallons of cider and beer. In 1761 the same thing happened. George Washington invited the entire delegation for cider the night before the election, and swept the board the following day. Cider ran the country. By 1767, per capita consumption of cider in Massachusetts was over half a gallon a day.

Cider had won them over. At Washington's estate at

Mount Vernon they distilled around 12,000 gallons of whisky a year, making it one of the largest distilleries in America. The distillery's ledgers indicate that Washington also produced apple brandy in 1798 and 1799. The orchards on his estate provided the apples for distilling. Hewes, Newtown Pippin and Roxbury Russet apples were some of the varieties used to produce the president's apple brandy.

Cider was big business: Colonel William Fitzhugh had 2,500 apple trees and made 9,600 gallons annually. On another estate owned by William Bray they were distilling and making Syder Royal. Distilling was all the rage until Prohibition in 1920s when orchards were set on fire and hard liquor was poured down the drain. (Even the Kennedys imported and sold liquor perfectly legally, before and after Prohibition.)

Thomas Jefferson was another fan of cider. He planted 1,031 fruit trees in Monticello's South Orchard. Cider-producing apples made up a large portion of this orchard, including Hewes crab apples and Taliaferro, 'the best cyder apple existing'. His wife Martha oversaw harvest and cider making. In the Chesapeake region the women often distilled the cider as well.

As a student at Harvard, John Adams (1735–1826) drank hard cider before breakfast. A gill to be precise – about a quarter of a pint. It did him no harm: he walked 5 miles a day and lived to the age of 90. When the ever wise and witty Benjamin Franklin overheard a young couple talking about the Garden of Eden, he said:

> It is indeed bad to eat apples, it's better to turn them all into cyder.

and

> He that drinks his cyder alone, let him catch his horse alone.

and

> Give me yesterday's bread, this day's flesh and last year's cider.

Another US president, William Henry Harrison (1773–1841), won the 1840 election on the 'log cabin and hard cider' campaign, aiming to recruit voters from backwoods hillbilly states. There is even a famous Harrison apple that was thought to be lost but was saved in the nick of time back in 1976. It gives a dry, robust cider. Henry Thoreau of Concord and Walden Pond, Massachusetts, would have approved. He wrote an essay called 'Wild Apples' in which he lamented the social changes and the apple's slow but sure fall from grace as it gave way to the temperance movement and German beer.

> The era of the Wild Apple will soon be past. It is a fruit which will probably become extinct in New England. You may still wander through old orchards of native fruit of great extent, which for the most part went to the cider-mill, now all gone to decay. I have heard of an orchard in

> a distant town, on the side of a hill, where the apples rolled down and lay four feet deep against a wall on the lower side, and this the owner cut down for fear they should be made into cider.

A contemporary American historian, the late William J. Rorabaugh, put it like this:

> In fact, so much cider was drunk that colonial Americans probably ingested more alcohol from that beverage than from their much more potent rum. And even with the increased popularity of distilled spirits after 1800, the annual per capita consumption of hard cider was 15 or more gallons. It continued until the 1830s to account for a significant portion of all the alcohol Americans imbibed. Hard cider disappeared only after the leaders of the temperance movement succeeded in persuading farmers to cut down their apple trees.

As the political barometer swung to the right, the cider apple would become a victim of its own success. Religion started to undermine its roots as the Puritan values of the seventeenth century re-emerged. In the 1920s Prohibition witch-hunts against alcohol were eerily like the Salem witch trials. Instead of a lynching there was a burning sensation not in the throat but in the orchard. Hardcore abolitionists set fire to many orchards and systematically destroyed cider making. Welcome to the Land of the Free. A very sad end to colonial cider. Washington and Jefferson wouldn't have stood by and allowed that to happen. Sadly, German beer then triumphed . . .

But a century later, things are looking up for cider. America has got its act together and cider is once more respectable and drinkable. From coast to coast, Americans can now write their own cider history. Every hillbilly American cider maker is a president-in-waiting, a novelist and poet.

7

SPARKLING CIDER

As the early colonists were boarding ships in the 1620s bound for America, clutching their bibles and saplings, something equally revolutionary was happening in the leafy shires of dear old sleepy England. Another new world was opening up with the work of skilled émigré glass makers. There was a distinct bubble of excitement which involved English cider makers pioneering a radical new approach to fermentation – the invention of sparkling cider that could be stored for several years within strong glass bottles. A deliberate, secondary fermentation with the addition of sugar which, when applied to certain French wines, is now known throughout the world as *méthode champenoise. Voilà.*

Some say this discovery by English cider makers was an accident, but it was in reality a series of experiments by some of the best minds in the country. Members of the Royal Society no less. There are many clues along the way: written accounts, letters and contemporary reportage, plus a fair bit of 'circumstantial'. It is an intriguing and compelling mystery worthy of the detective skills of Inspector Poirot, who would need all his little grey cells to examine the evidence: the paper trail, dark glass and the decidedly English habit of bottling it all up.

During the seventeenth century there were many severe winters. A mini Ice Age. The Thames froze over for months. There were ice fairs. Prolonged frosts with wet summers made English vineyards unsustainable. Mildew took its toll on vines, so large houses expanded their orchards and ramped up cider making. Wine also suffered. Cromwell's Navigation Act 1651 was protectionist and prohibited Dutch importation of wines from the Rhine, Moselle and much of France. On the top table, good wine was in short supply. There was a desperate need to fill the gap.

So the English upper crust decided to solve their own problems in their own way. Necessity – the mother of invention. Cider makers rose to the challenge and began working behind the scenes. Why not make fine cider of almost equal strength to wine? Not a bad idea. Cider made from Redstreak, Foxwhelp and Golden Harvey apples were reckoned by Thomas Andrew Knight to be about 11 per cent abv.

Cider became the national drink, something Parliamentarians and Royalists actually agreed on. Orchards and cider were good news and *bottled* cider was the way forward. Aristocrats and gentry, having invested in the American colonies with mixed success, now wanted a business nearer to home that was more lucrative. Glass making was just the ticket.

The most exciting innnovation was the production of

fine sparkling cider. Towards the end of fermentation in wooden barrels, if you judged it right and the bottles didn't break, you bottled your cider and ended up with a naturally sparkling cider, a technique known in the trade as *méthode ancestral*, *pétillant naturel*, or '*pét nat*'. But you have to have very strong bottles. Cider makers went one step further and added sugar to the cider to give it more fizz. If you used too much sugar, the bottle became a homemade bomb and produced what they called 'Pot Gun' cider.

This technique was only made possible by the discovery of newfangled dark green, almost black bottles which emerged from glass houses, *c.*1615, in Vauxhall and Broad Street in London as well as the Forest of Dean, the Sussex Weald and Purbeck. Many Huguenots had arrived during the reign of Queen Elizabeth I and joined Italian glass makers who had secretly escaped from the island of Murano and the Venetian Republic.

So, as one group of emigrants sailed off into the sunset towards America another group of immigrants came in by the back door from Europe. Glass was big business and secrecy was paramount. Glass makers like Jacob Verzelini, Jean Carré, Isaac and Peter de Bongard, Paul Tysack and Anthony Voyden were suddenly in demand. Techniques and patents were jealously guarded; skullduggery was commonplace. English glass entrepreneurs like Sir Jerome Bowe and Edward La Zouche soon emerged.The Huguenots still worked the furnaces, but their names slowly became Anglicized. Jean Carré from Arras became John Carr, Pierre de Bongard became Peter Bungard and Balthazar de Henesell became Hensey.

The only problem was that glass kilns used vast quantities of precious firewood to get high temperatures, *c.*1500° Centigrade. Iron makers also needed firewood and charcoal to smelt iron ore and the navy needed timber to build ships in naval dockyards. It took about 2,500 oak trees to construct a single warship. Competition was fierce. When chased out of towns glass makers became itinerant and often went to live in remote forests.

Another problem was that certain estates sold timber too soon and too young. Around Petworth in Sussex huge swathes of timber were cut down just to pay gambling debts. Even in Elizabeth I's day, Parliament realized that there was a finite amount of timber in the country. The upshot was a Royal Proclamation which forbade anyone from using wood for heating glass kilns. The language is interesting. The transcribed document exists in a book called *Old English Glasses* by Albert Hartshorne (1897). Just imagine that you are James I, decked out in ermine robes, with trumpets, fanfares and peacocks on hand.

Royal Proclamation No. 42 –
issued from Greenwich on 23rd May 1615.
'A Proclamation touching glass'

From the day of the date of this Our Proclamation, no persons or persons whatsoever shal melt, make, or cause to be melted or made, or any kind, forme or fashion of Glass or Glasses whatsoever, with Timber or wood, or any fewell made from wood within this Our Kingdom of England and Dominion of Wales . . .

There were severe punishments for infringement. Coal was king. James I also banned the importation of glass bottles from France. Woods were earmarked for shipbuilding. The Proclamation was designed to protect his timber racket supplying the navy but ended up as a conservation measure to protect forests. It forced the iron makers to turn to coal and inadvertently kick-started the industrial revolution by turning up the heat. A century later coal-fired steam engines emerged triumphant.

But what evolved in the glass world was something very different from the fashionable, delicate, ornate, Venetian glass goblets. This glass was tough and useful. Dark, strong and very English.

Proclamation No. 42 was rigorously enforced but the Huguenot glass makers like Paul Tyzack and Abraham Liscourt objected, saying that they had been experimenting with coal for a good number of years in the Forest of Dean. What they found was that impurities within coal, i.e. iron, sulphur, manganese and potassium, had combined with the sand, potash, lime and alumina at much higher temperatures. At Kimmeridge in Purbeck, Dorset *c.*1618, there were terrible reports of black noxious fumes created by burning oil shale.

At first the glass makers discarded the dark glass which resulted. Although not beautiful, it was very strong and much more useful. Strong, dark green glass bottles were just the ticket. Something very important had indeed happened between 1600 and 1620. But it was never plain sailing. Huguenots were notoriously secretive and knowledge was only shared between partners or family members.

In 1615 a naval admiral called Sir Robert Mansell acquired the outright monopoly on glass making which, with one or two hiccups, lasted till 1642.

For centuries bottles had been fragile. They were used in Elizabethan times to take wine, beer or cider from the barrel to the table with a cork or wooden stopper in them. Nothing grand: just a convenient way of serving it without the need for a jug. Bottles were surrounded by rafia if they came from Italy, like Chianti, or small wicker baskets, strong and flexible. There was no reliable way to safely store or transport wine in glass bottles. That was still done in barrels. But now, with strong bottle glass, everything changed.

Seeking Barrilla

Glass making was a complicated and secretive business. Many were in awe of the process which had seemingly magical properties. Artisan skills turned simple raw materials into 'Objects of Great Wonder'. One man who gives a real flavour of these exciting times was James Howell (1594–1666), an Oxford linguist and man of letters. His tutor's uncle was Sir Robert Mansell who employed him as manager of Broadstreet Glass Works in London. Warm work. Howell admitted he couldn't take the heat and 'should in a short time hath melted away amongst those hot Venetians'. In 1618 he became Mansell's roving Continental agent, bringing back glass makers with specialist knowledge from Italy, and, from Spain, *barrilla*, a salt-tolerant shrub found on Mediterranean shores.

The observant and enthusiastic Howell wrote many letters. From Venice he wrote to his brother on 1 June 1621: 'The Art of Glassmaking is here very highly valued for whosoever be of that Profession is a gentleman *ipso facto* and it is not without reason it being a rare kind of Knowledge and Chemistry to transmute Dust and Sand (for they are the only ingredients) to such a diaphanous pellucid dainty Body as you see a Crystal-Glass is.' The purest and best raw materials available. Sand and lime. Glass and class.

Howell then wrote to Christopher Jones, Gray's Inn, about *barrilla*: 'I am to send hence a commodity called Barilla (saltwort much used in Catalan glassmaking) to Sir Robert Mansell for making of crystal glass and I have been treated with Signior Andreotti, a Genoa merchant, for a good parcel of it, to the value of £2000 a vast sum by letters of credit from Master Richant and upon this credit might have many thousands more, he is so well known in the Kingdom of Valencia.' No wonder fine drinking glasses like the Scudamore flute were so expensive.

The best Spanish *barrilla* – prepared by master *barrilleros* – contained about 30 per cent Na2CO3. Sodium carbonate acts as a flux for silica, lowering the melting point of the mixture. In 1877 Kingzett described the importance of the *barrilla* trade to Spain as follows: 'So highly was the product valued, and the importance of the trade regarded, that by the laws of Spain the exportation of the seed was an offence punishable by death.'

Sources of soda ash and methods of manufacture were

secrets that were zealously guarded for centuries. James Howell knew what he was doing and was a trusty pair of hands. Ornate crystal glass is one thing, but dirty, dark green bottle glass was something else. The knowledge Howell picked up about glass making abroad and in London put him in a very good position to give advice to Sir Kenelm Digby when they met in Madrid in 1623.

Howell was in Madrid on a diplomatic mission to secure the release of an impounded British vessel. Sir Kenelm was on another diplomatic mission (wearing false beards) with the young Prince Charles to court the Infanta. These events led indirectly to glassmaking via sword fighting. Howell met Digby over treatment of wounds to his hand sustained in separating two friends who were duelling. By coincidence Digby had also been wounded on a moonlit night when being serenaded by a young lady at her window in her nightdress. It was a honey trap. Digby and his four companions were set upon by fifteen upmarket Spanish desperados. They were both wounded and *hors de combat*. They bonded over duelling. *Touché. En Garde.*

Four days later when his hand was badly infected Howell visited Sir Kenelm and asked to be cured by his 'sympathetic powder', as the surgeon intended to amputate the hand. Sir Kenelm obliged and soaked his blood-stained garter in a bowl of water and green vitriol crystals. (Green vitriol was very useful. Mixed with tannic acid it also made oak gall ink for manuscripts.) The pain disappeared even though Howell was on the other side of the room. Sympathetic magic or an early case of long-range antibiotic?

A good story, though, as the cure was probably brought about by washing and bandaging the wound.

James Howell and Sir Kenelm then became 'very intimate' for the rest of their lives. Similar in spirit, witty, sharp-sworded and ambitious, bonding not just over duelling but over the inner secrets of glass-making. Vital for Digby when he later 'invented' the shape of the wine bottle. Seeking Barrilla paved the way.

Howell later wrote a guidebook to the Continent, *Instructions for Forreine Travel* (1642) and a book of letters when he was in Fleet Prison for eight years. But prison was no disgrace for a writer: even Digby had been in prison for killing a man in a duel, doing time in the 1640s at Winchester House, Southwark, next to the famous glassworks. Admiral Mansell had spent only two weeks in prison for a massive fraud. Some people get all the luck. Glass was indeed class. Sir Kenelm knew all about that.

Sir Kenelm Digby

Today we take the standard shape of wine bottles for granted but they evolved slowly and went through several stages: small, squat, tall, thin, bulbous, cylindrical, onion-shaped, squashed, square. Glass is malleable and fluid even when you think it is solid. When in the furnace it changes form with the long breath of the glass maker, and as he blows each contour of the neck is elongated, the body firm and round. Beauty and fragility, utility and strength.

Sadly, very few early dark green bottles survive intact. Even fragments are rare. Broken shards, called cullet, were often added to the next firing.

The first dark green wine bottles were shaped with a bulbous bottom and a punt, or kick, the indentation that makes the bottle much stronger. Then a long neck and the all-important string lip or rim about an inch below the top, used for tying down a cork. Champagne bottles have rims to this day, their corks secured with fine wire. You may have bought similar bottles in France without realizing their intriguing history.

But once the liquid is within the bottle, whether wine or cider, it can be stored for years and so transform itself into something of great beauty and depth. What is vital is that bottle glass can take the pressure which builds up when a secondary fermentation takes place. Fermentations are powerful things. By the 1660s bottles had become squatter with shorter necks, a bit like large hand grenades.

These newfangled dark green bottles could stand a bit of punishment. They were just what cider makers like Lord Scudamore from Holme Lacy, near Hereford, needed to keep their cider in, not just for weeks but for years. Word on the street was that Sir Kenelm Digby, courtier, natural philosopher, poet, pirate and Fellow of the Royal Society had a key role in inventing the wine bottle and perfecting a design that could be replicated. At least that is how the law saw it.

In 1662 the legal status of the 'new' bottle was in dispute and ended up on the Attorney General's desk. Glass making was fraught with legal problems during this period, not least patents, monopolies and quirks of distribution. The long and short of it was that on 1 September 1661 John Colnett, a Huguenot glass maker, whose original, Belgian name was Jean Colinet, received letters patent from Charles II to make glass 'Gallons, Pottles, Quarts, pints and half pints'. He claimed he had 'first invented, made and attained to ye perfection of making glass bottles' and that during 'the late rebellious times' others had appropriated his invention. Parliament agreed and gave John Colnett what he wanted.

However, there were rumblings in back alleys. Four well-known London glass makers, John Vinion, Robert Ward, Edward Perceval and William Sadler, presented a petition countering Colnett's patent. They claimed that Sir Kenelm Digby was the 'first' inventor of the English bottle and that he had employed them with Colnett to make the bottles 'neer thirty years since', which winds the clock back to around 1631–2.

They said that Digby had later abandoned his interest and that the trade 'had been of publique use at several glasse houses in England for many years'. No surprise there. In 1633 Digby had become a recluse in his laboratory in Gresham's College and had been dabbling in scientific experiments since 1624, with the odd spell of piracy in between.

Ultimately Sir Geoffrey Palmer, Attorney-General and a no-nonsense Royalist, took the side of the four glass makers. The patent was dismissed. Sir Kenelm was now the official 'inventor' of the wine bottle! He could possibly have had a licence from Mansell, whom he knew through the navy, to run a glassworks at Newnham on Severn. There had been a glasshouse there since 1613 initially run by Rowland Ferrice. Mansell took it over in 1616. A wonderful spot beside the river with coal mines and cider nearby in the Forest of Dean. This possibility was first raised by Eleanor Godfrey back in 1975. Digby also had the sealing wax monopoly in Wales and on the Welsh borders. Sealing wax was another important part of the bottling process as it sealed the cork and thus stopped the air or any moisture getting in. Wax is still used today.

But who exactly was Sir Kenelm Digby, the 'Ornament of All England' and one of the most interesting men of his age? Digby (1603–65) was born at Gayhurst House, Milton Keynes. His father Sir Everard had helped Guy Fawkes and his gang in their bid to blow up the Houses of Parliament in the Gunpowder Plot and was publicly executed for the part he played. Hanged, drawn and quartered. Digby was only 3 years old at the time.

When he was 15, Digby went to Gloucester Hall, Oxford, where he was taught by Thomas Allen, a friend of the Scudamores. Mathematics, manuscripts and alchemy. He spent three years on the Continent between 1620 and 1623, where Marie de' Medici fell madly in love with him (or so he later said). She was thirty years older than him and Digby only just escaped with his life and reputation intact. Marie de' Medici later became Charles I's mother-in-law, following Charles's marriage to her daughter Henrietta Maria. Small world.

Digby then fled to Angers, and spread the rumour that he was dead. An old ploy. Then Florence for two years. He survived smallpox, became fluent in Italian, collected books and manuscripts. Addressed the Accademia dei filomati three times in Siena; and an aged Carmelite monk gave him the recipe for his famous 'powder of sympathy'. He also met Galileo. Why not?

Digby returned to Madrid where his uncle was ambassador, which was when he accompanied the young Prince Charles in disguise on the courting escapade that failed. Back home, Digby was very taken with the great beauty and socialite Venetia Stanley of the Percy family, whom he had known since childhood. When Digby faked his death, Venetia got to hear about it and was so upset that she played the field back in London.

When he returned in 1625 the pair married secretly. But just as he was about to settle down, he became a privateer and knobbled a few ships off Scanderoon (modern Iskenderun, Turkey) in the eastern Mediterranean. He captured Spanish and Flemish vessels, French, Venetian and Dutch. He wasn't very choosy so this caused a diplomatic rumpus. Reprisals were taken against English merchants in Aleppo. Digby made some money, a few friends and a few enemies. The Digbys then built a house in Aldwych Close, Holborn. Privateering paid off: he got a job with the navy. His sealing wax monopoly was extended to Northumberland and Cheshire. Stamping legal documents and letters was a lucrative business.

Sadly, on May Day 1633 Venetia died very suddenly from a brain haemorrhage. Famously she was painted by Van Dyck on her death bed. Sir Kenelm mourned her deeply. Then he renounced the world and became a hermit alchemist at Gresham's College in Holborn.

Fig. 7.1 'Sir Kenelm Digby. Diplomat, courtier, naval administrator, alchemist, pirate, glassmaker, cidermaker, food and drink historian', 1654

Digby's laboratory was stuffed with chemical apparatus. An inventory of his workshops in August 1648 included a vast array of glassware, retorts, ovens, calciners, furnaces, twenty lymbeck (distilling equipment) and many bottles. He was a scientist and natural philosopher, with an enquiring mind, a pirate and author of *Loose Phantasies* and *In Praise of Venetia*. Above all he had time and intelligence to experiment with glass . . .

Digby was also up to his eyes in the old manuscripts he had inherited from Thomas Allen, which he gave to the Bodleian Library, Oxford, more than 200 from Allen, 100 of his own plus thirty-six Arab manuscripts in the 1640s, which triggers another line of thought. Was it not the alchemist Kabir Jabir ibn Hayyan who in the eighth century AD wrote a treatise on glass making, the *Kitab al-Durra al-Maknuna* (*The Book of the Hidden Pearl*), and used marcasite, called 'white iron pyrite', i.e. iron sulfide. Kabir also used manganese dioxide in glass making, to counteract the green tinge produced by iron – which is still used today. Digby needed green glass and strength. They were on the right path but at different times and in different countries. Digby valued strength over clarity.

As for colours in glass: 'For blue: oxyde of cobalt. Green: oxdye of iron or copper. Violet: oxyde of manganese. Red: oxydes of copper and iron. Purple: oxyde of gold. White: oxyde of arsenic and zinc. Black: oxyde of manganese, cobalt and iron . . .'

Copperas, alias green vitriol, was made from iron pyrites, often referred to as fool's gold and found mainly in coal measures. So the green vitriol could have occurred as an impurity in the coal or in the river sand at Newnham on Severn which was used for glass making.

At court, Digby had a good friend in Sir John Winter through Queen Henrietta Maria. He was her chancellor and Sir John her secretary. Sir John had many interests in the Forest of Dean including ironworking and coal at Newent and Newnham on Severn, as well as glass making. The Huguenots must have used his coal to make glass when the wood ran out. All signs point to Digby making glass at Newnham, away from the prying eyes of other London glasshouses. But alas no hard evidence.

Digby also made cider and collected recipes. His book

The CLOSET of Sir Kenelm Digby OPENED contains his experiments with making various drinks. A gourmet's delight. When he makes cider he mentions bottling cider, secondary or malolactic fermentation when the weather gets warmer and 'naughty little dregs' - i.e. yeast at the bottom which reactivates the fermentation and thus breaks the bottles.

We may never know exactly what Sir Kenelm Digby was up to with glass in his laboratory in the 1630s, but he was certainly going in the right direction, and was in the right frame of mind to invent wine bottles. Right place, right time, right intellect. The Attorney General thought as much when he awarded the four glassmakers the patent in 1662. Cider was now bottled up and ready for action. More importantly, Digby did not invent champagne as some wine writers surmise.

Court cider

Imagine a 'sweet especial rural scene', Herefordshire, *c.*1630, a few miles from Hereford. Take the winding road past Sink Green Farm with commanding views of the Wye. Newly introduced red and white cattle graze contentedly amid orchards. There is a rambling Elizabethan mansion with outhouses smelling of apples and cider making, and a massive perry pear tree groaning under the weight of fruit down by the river.

Lord Scudamore (1601–71) attended Magdalen College, Oxford, then Middle Temple, before visiting France for a year or two to whet his whistle. In 1615 he had married Elizabeth Porter and acquired her father's estates and orchards in Monmouthshire and Gloucester. Scudamore also inherited his grandfather's estate with 13,600 acres and a further 1,840 acres in Worcestershire. All with considerable orchards. Scudamore was MP for Hereford until 1629 when Charles I decided to rule without Parliament, much as his father, James I had done.

There was no jousting before lunch, but Lord Scudamore worked his way into court with *'rondlettes of syder'* (small barrels of 18 gallons), judiciously gifted to impress the courtiers – men like George Villiers, the Duke of Buckingham. It seemed to work. Flattery or bribery matters not, his cider spoke volumes.

Fig. 7.2 'Lord John Scudamore. MP, diplomat and cidermaker' (1601–1671)

In 1627 Scudamore gave two gold cups to Thomas Coventry, 1st Baron Coventry (1578–1640), lawyer, judge and Lord Keeper of the Great Seal of England. Then more *rondlettes* of Holme Lacy *syder.* These were his calling cards, a way of oiling the wheels at court. Lord Scudamore must have been very confident that his cider would do the trick. At court you are only as good as your cider. *Plus ça change*?

First Rondlette: to Sir Thomas Edmondes, Treasurer of the Household (1563–1639), diplomat and politician who served three monarchs, Elizabeth I, James I and Charles I, as well as Treasurer of the Royal Household. His father had been customer of Plymouth which meant he had control of most of the Cornish ports. He would have known all about Drake, Devon cider and pirates.

Rondlette 2: to George Villiers, 1st Duke of Buckingham (1592–1628), the king's favourite, close to Charles in more senses than one. James Howell said he was 'on his way up' and Sir Kenelm Digby and Lord Scudamore knew him well. Villiers was later stabbed to death in the Greyhound, a Portsmouth pub locally known as the Spotted Dog.

Rondlette 3: Sir Roger Palmer (1577–1657), Master of the Household. Hands-on below stairs with servants, meals, etc. Food and drink master of etiquette and victualling. A wise move.

Rondlette 4: Edward Somerset, 4th Earl of Worcester (1550–1628), House of Beaufort. A big cheese. He had fifteen children. One of his daughters, Lady Blanche, was very courageous. She married Thomas Arundel of Wardour

Castle, Wiltshire, and, during the Civil War, after her husband was killed, she was besieged in the castle. She held out for six days with just twenty-five men, her children and maid servants against a Parliamentary force of 1,300 who pounded the castle with artillery. Cider would have come in handy. I know the castle. We used to have cider parties there on moonlit nights and climb the crumbling walls.

Rondlette 5: to Archbishop William Laud (1573–1645), also close to Charles I. Bishop of Bath and Wells then Archbishop of Canterbury. Another great friend of Scudamore's. Religion was a minefield. Laud was not liked by the Puritans because he advocated rule by bishops. He was later imprisoned by Parliamentary Rule, tried for treason and executed.

So, of the five people who received rondlettes of cider, two met sticky ends. Scudamore was close to Buckingham and Laud. Their deaths affected him greatly. During the Civil War, Scudamore was captured and imprisoned in London. When he went back to Herefordshire, he kept his head down in Holme Lacy and perfected his cider making. Cider philosophy was about to enter a very interesting phase.

Scudamore's flute

Imagine Lord Scudamore takes a train from Hereford down to Newport, then catches the fast train to Paddington, then the Tube to Barbican on the Circle Line and then a five-minute walk to the Museum of London. St Paul's is only a stone's throw away. And there, deep inside the museum, displayed on one of the dining tables laid out for a small banquet, Lord Scudamore can see one of his very own tall, slender drinking glasses. On one side is Charles I's crest, on the other Scudamore's own crest. In the half-darkness, you could easily miss it. The glass is 14 inches high, tapered, elegant, extravagant, like a champagne flute. Only it is not a champagne flute. Sparkling champagne, as we know it today, had not been invented then. Not by a long chalk.

Its creation would have required the sort of glass-making techniques that James Howell was researching in Venice, though more likely to have been made by Venetians in Antwerp or London. The Scudamore flute, also known as the Chesterfield flute, is rather unwieldly and more for making a statement perhaps? So what had Lord Scudamore been doing to warrant such a glass? And why was he so keen to display his Redstreak apples also engraved on the glass?

When Lord Scudamore was not in London or Paris, he was on his estate in Holme Lacy. Luckily for us, household accounts for 1631–2 have survived and very interesting they are, too:

> Paid Jeffrey Cook for carrying 6 hogsheads of sydar to London 1 6 6*d*.
> Paid Jeffrey Cook for bringing downe from London a dozen and a half bottles 0 4 6*d*.
> Paid Henry Prosser for making 6 stooles 0 1 0*d*.
> Paid Wilcox for 2 dozen and half quart bottles 0 7 6*d*.

Thrid [thread] and incle[2] for them	0 6*d.*
1 dozen and a half pint bottles	0 3 9*d.*
6 dozen corkes	3 0*d.*
a basket corde and porters carryge	1 3*d.*
a watering pot	3 6*d.*
a great knife to cut bread	1 6*d.*
a new lock for sydar house doore	0 1 0*d.*

This is just the sort of kit you would need if you were experimenting with bottling cider. Scudamore was known to have 'rare contrived sellers in his park for keeping cider with spring water running into them'. He also had a lake and an ice house. I have seen both the lake and the ruins of the ice house. 'Stooles' refers to racks or low shelves. The reference to bottles brought down from London is very interesting. Wilcox was a glass maker in Gloucester, possibly Newnham on Severn as well. It seems well-organized and planned, as if something new and exciting was being undertaken. 'New lock for sydar house doore' rather clinches it. Lord Scudamore knew exactly what he was doing.

Then Scudamore was appointed Charles I's ambassador to Paris in 1635. His cider had done the trick! A plum job at last. Bottled cider back home would have to wait.

Louis XIII (1601–43) was King of France and Navarre, elder brother of Henrietta Maria, Charles I's queen. As a Catholic, Henrietta Maria did not support Scudamore. A complicated scenario. Louis's mother was Marie de' Medici,

2 'Incle' is linen tape or yarn used for holding the corks down and binding them tight. A vital part of the operation.

who had been after Sir Kenelm Digby when he was a young blade in Madrid. What a tangled web.

Sadly, Scudamore's time in Paris was marred by Charles I's quixotic foreign policy wavering between supporting France then Spain. In April 1636 Scudamore, who was pro-Spain, was outmanoeuvred by the appointment of Robert Sidney, Earl of Leicester, who was more adept and wilier – a hardcore Protestant.

However, Scudamore did not waste his time in Paris. Far from it. He was there at his own expense, which was considerable. Paris was the hub of intellectual ideas and philosophy. It was here that Scudamore would meet Digby in 1635. He also met the philosopher Thomas Hobbes. But Scudamore felt humiliated and outmanoeuvred not just by the French, but by Leicester as well. He missed his wife, his cider, his cattle and Herefordshire and asked to be recalled. After four years in Paris he returned home in March 1639.

His household records for 1639 show some additional accounting of a cidery nature as he gears up for cider production:

Gathering grafts for London	8*d*. per diem 7 days
Mending the cider mill	1*s*. per diem
Apple gathering	6*d*. per diem
A cooper	1*s*. per diem
A common labourer	6*d* per diem
Making a hogshead	5*s*.
Apples and cider sent to London	2,600 at 6*s*. the 100 £5. 16*s*.
Pears carried to London	*+ 6 bottles of cider*

By the autumn of 1639 Lord Scudamore was taking the fruits of Holme Lacy to London.That he took six bottles of cider is remarkable. It meant his experiments with bottling in 1632 had paid off.We may never know how many bottles broke, or what happened in 1633 and 1634, but he was confident enough to take bottles up to London. It was a rough old ride in a cart, with the cider packed in straw and wicker hampers. The cider must have been very good. But the real question is whether or not the cider was sparkling when opened at the other end. If he was using sugar there was a risk that bottles would explode, leaving the recipient and Scudamore's reputation rather damp and showered in glass.

At Christmas 1639 Lord Scudamore had open house at *Hom Lacie* for nineteen days.Anybody who was anybody in Herefordshire was invited.The records of food and drink consumed are in the British Museum:

> Oxen, cowes, Muttons, bacons, Pigs,Turkeys, geese, capons, Partridges, pheasants, grouse, eeles, carps, samons, coddes, oysters – 300, pickled oysters (3 barrels) Olde Lyngem Cheeses, Candles 359, Eggs 1017, Blackbirds 2 dozen, Peacocks 3, Calves haggases,Tripes, udders, Calves' heads, calves feet, Neates tongue
> Beere 54 Hogsheads at 10*s*. ye hogshead
> Cider 17 Hogsheads at 17*s*. ye hogshead
> Aile at 15*s*. ye hogshead
> + olives, Cape, samphire, nutmeg, dates almons cloves, mace, cinnamone, rice, Sugar loaf, sugar powder, Currans,

pruens, raysons, Claret, Sacke, Muscadine and white wine as well as Musicke of Hereford, Welsh Harper, Blinde Harper, Organiste, Singinge Boys.
In all: Cellar was £32/17 – Wine £44.14-1
Grand total. £233/8/6 + Gifts valued at £131 3*s.* 2½*d.*

That sort of party would have bankrupted most houses of the day, but Scudamore had been away for nearly four years. He needed to keep his presence buoyant in his own backyard. But Civil War was looming. If Scudamore thought France was bad, far worse was in store for him at home. He was not a military man and it showed. Scudamore was indecisive. He spent more time squabbling with other Royalists than preparing for action. In April 1643 Hereford surrendered almost without a fight. Scudamore was captured and taken to London where he lived under house arrest for four years.

All of Scudamore's money, plates, household goods and furniture were seized and sold. His papers disappeared. His losses were estimated to be around £22,190, today about £2.5 million. Not a good outcome if you were a cider maker.

This is where the Scudamore flute comes in. Was it purloined, confiscated, sold off or hidden away in some priest's hole? It was supposedly made around 1650 and the fact that it survived at all is remarkable. Yet I find the date perplexing. I think it is earlier, much more likely to be 1635, just before Scudamore went to Paris as ambassador, so that he had a set of interesting glasses with which to entertain his guests. The royal crest speaks volumes. By 1650 Charles no longer had his head on his shoulders. The flute has to be playing a tune before the Civil War.

The Scudamore flute is magnificent. Not just coats of arms but roses, fruit and flowers, a stag beside a gate and five trees, Redstreaks and the letter S (for Scudamore) repeated three times. The very thing for an aspiring ambassador.

The flute and its companions were for drinking sparkling cider, and this in an age when sparkling champagne did not exist . . . not even in France. Imagine that. These glasses were commissioned by a man on the way up, stepping out into the diplomatic world, not someone who had been imprisoned for four years, lost much of his wealth and had Parliament watching his every move.

Scudamore's flute is proof positive that English upper classes were drinking sparkling cider at a much earlier date than anyone has suspected. Possibly even the 1630s, which means that they knew exactly what they were doing with bottles and secondary fermentation. In those days cider was called 'vinous' if it was of very high quality, close to wine. The two words wine and cider were often synonymous at the top table. His cider was called 'vin de Scudamore' by an Italian prince when visiting Oxford.

According to Revd John Beale, a Fellow of the Royal Society, Lord Scudamore turned cider from 'an unreguarded windy drinke fit only for Clownes and day labourers into a drink fit for Kings, Princes and Lords'.

Fig. 7.3 'Lord Scudamore's flute also known as the Chesterfield flute', A soda cristallo drinking glass, 14 inches high, *c.*1640

Talking trees

We need far more information about these new techniques of bottling cider to determine if the cider actually sparkled. Sadly, Lord Scudamore left very little in writing that has survived. A cellar notebook would come in very handy, but the Civil War intervened. We only have Scudamore's flute and his cider's reputation. In other circumstances

Scudamore could have turned his experiments into a business, but he preferred religion, cattle and orchards as his constant companions. Why spoil a fantastic hobby by turning it into work? Yet he was pushing at the boundaries of knowledge. He encouraged many people to plant orchards with Redstreak cider apples and make the shires fruitful and economic.

But *sparkling* cider? It is all in the method: trial, error and careful observation over many years. Patient skill, a deep pocket, quiet determination and a strong belief in your own philosophy. These Scudamore had in abundance. When the Civil War ended in 1651, he retired to his estates, but he was by no means the only cider maker experimenting with bottling.

Next in the roll call is Ralph Austen (1612–76), an ardent Puritan with a small cider factory in Queen Street in the heart of Oxford. Austen was a very different kettle of fish from Lord Scudamore. He came from Leek in Staffordshire. His mother was a cousin of Henry Ireton, a Roundhead general who married Oliver Cromwell's daughter Bridget. Austen was a strict Parliamentarian. After the siege of Oxford in 1646 he lived there for the next thirty years.

Austen worked closely with the military authorities and became registrar to the Parliamentary visitation of Oxford University. A bit like the Thought Police. He was unpopular with the Royalists, which is why, later on, he was excluded from the Royal Society. The cider factory was in his large walled garden off Queen Street. Austen also ran a large tree nursery and supplied 1,000 apple trees for the city solicitor and then sent out more than 20,000 plants which, at 1*s*. a tree, brought him in £1,000 a year. A very tidy sum in those days, but like many businessmen with grand ideas, he was always in debt. He employed two workmen and was helped by his wife Sarah. Austen was always badgering Cromwell to make the planting of orchards compulsory.

Fig. 7.4 'Walled garden with fruit trees', from *A Treatise of Fruit Trees*, by Ralph Austen, 1657

Ralph Austen not only made cider, he wrote books: *A Treatise on Fruit Trees* (1653) which was dedicated to Samuel Hartlib with a section on cider and perry. The book was ready by 1651 but it took two years to get funds together to go to print. Austen believed that cider would promote health, allay sedition and eliminate the call for imported wine. A second edition appeared in 1657 with *The Spiritual Use of an Orchard*, where trees talked to each other and the gardener.

Being an ardent Puritan, Austen was a bit too religious for most people, even for those times, but his bottle-fermented cider and apple trees were much in demand. He made cider till the day he died. Eventually word reached Cambridge when Sir Isaac Newton requested grafts of his apple trees and seedlings. Fame indeed. As for the all-important art of bottling cider and how to keep cider in good condition, Ralph Austen says this in 1653:

> Cider maybe kept perfect a good many years, if being settled it be drawn into bottles and well stopt with corkes and hard wax melted thereon, and bound down with pack thread, and then sunke down into a well or poole, or buried in the ground, or sand laid in a cellar.

Crucially, an additional note was printed in the margin of the 1657 edition: 'Put into each bottle a lump or two of hard sugar or sugar bruised.' He was on the right track. When Austen says, 'Cider maybe kept a good many years', this could be anything between five to ten years, referring either to his own cider or to Royalist cider made before the Civil War. Cider that had to be abandoned or hidden in haste down wells. That takes it back to 1650 or even the early 1640s.

When Austen says 'settled' he means that the primary fermentation has finished. This is crucial. Full-blown fermentation creates great pressure as the carbon dioxide builds up in the shape of small bubbles that eventually shatter bottles as they explode. Dangerous and expensive. You needed strong bottle glass that could withstand high pressure, which is why Huguenot glass makers were so important. You also needed good corks and a secure way of tying them down.

The point of settling the bottles in the sand is that if one bottle explodes its neighbours will not be affected. Austen is careful to advise a cool, stable place such as a cellar. Lord Scudamore had a stream of water passing through his cellar to keep the cider at a more or less constant temperature. Bottles break if the glass has hidden flaws.

Austen's reference of 1657 is the first printed mention of the deliberate addition of sugar to a bottle of cider, or even wine for that matter. In those days cider was often regarded as a wine. The addition of sugar to a finished cider or wine was a fundamental part of what is now called *méthode champenoise*. But Austen's account predates any French claims by almost forty years.

Even if it was *méthode ancestrale* or '*pét nat*', the continuation of the first fermentation in bottle is still a real step forward and would have produced a fine sparkle, akin to Normandy *cidre bouché*.

As for France, Dom Pierre Pérignon, who is often credited with the invention of champagne, was only born in December 1638. He was only 1 year old when Lord

Scudamore was sending bottles of cider to London and only 14 or 15 when Ralph Austen's book on cider was first published. Dom Pérignon did not enter his monastery till 1664. Time and ingenuity were on the English cider makers' side.

As for Austen's 'talking trees', *The Spiritual Use of an Orchard* contains several interesting quotes:

> Fruit trees discover many things of God and many things of ourselves, and concerning our duty to God. We enquire of and discourse with Fruit trees when we consider, and meditate of them, and search out their virtues and perfections which God hath put into them, when we pry into their natures and properties, *that is speaking to them*.

A dialogue with God through the trees. Whether it be the Christian God or Cider God matters not. As to good health:

> Cider is more conducing to health and long-life then Beere and Ale (though there are also good liquors, especially for some persons) for cider is a cleare Liquor without dregs, and does not onely not leave any dreggs in the body, of its own substance, but it hath the property to cleanse the body, & carry downe superfluities and hurtful humours in the body, which other liquors and meats have engendered, and left in the body, which are the seas of many distempers and diseases.

Thus cider promoted good health and soothed the mind whereas beer provoked sedition. Austen was also in correspondence with Samuel Hartlib and these letters leave a paper trail. Austen propagated some seventy-five varieties of fruit-bearing trees, bushes and vines, and sowed thousands of timber trees in 1656 to avert a national shortage of wood.

Ralph Austen was a radical Green at heart. He had his entire house painted green inside, apart from the best chamber, which was reserved for a wealthy lodger. His nursery extended to 27 acres. When he died he had seven hogsheads and 528 bottles of cider in his store as well as a fully equipped 'Sider House'.

The Hartlib Circle

Ralph Austen was not the only man writing about bottled cider and storing it for a number of years. Cider was discussed at length within a group called the Hartlib Circle, started in the 1630s by Samuel Hartlib (1600–62), the 'Great Intelligencer of Europe'. Of German origin, Hartlib was born in Poland, his father a wealthy Baltic merchant, his mother the daughter of an English merchant in Danzig. Hartlib went to Cambridge and lived in London in Axe Yard, close to Downing Street. Samuel Pepys was a neighbour and he mentions *syder* seven times in his diaries including bottles of *syder* and 'brave French *syder*'.

Hartlib was a walking, talking, one-man university with a vast array of contacts. Widely respected, he had tentacles stretching deep into European thought and experiment.

Letters from 400 correspondents range from subjects such as natural philosophy, seed drills, calculators and siege machines to orchards and cider making. In the 1650s the Herefordshire clergyman John Beale wrote to Hartlib roughly once a week outlining his latest observations on cider. Buried deep within these letters I found the earliest accounts of bottle-fermented, or 'mantling', cider.

In 1662, just when the Royal Society was taking off, Samuel Hartlib died impoverished. After his death the remaining diaries, letters and manuscripts were bought by William Brereton (1611–64), who lived in Cheshire. Brereton died two years later. John Worthington, an academic and past Master of Jesus College, Cambridge, found the papers. The letters then disappeared altogether for more than 250 years, until 1933. Remarkably they were found in a solicitor's office by George Turnbull of Sheffield University and have now been transcribed and catalogued.

Samuel Hartlib's correspondence runs to over 25,000 folios of original material and are a staggering insight into people's minds and thoughts at that time. Within this collection are seventy letters related to cider, mostly from Revd John Beale. They show an almost hyperactive mind, dipping from one subject to another, always pithy and accurate, with many anecdotes and testimonials. Page after page of observations. The postal system must have been quite good between Hereford, Oxford and London

The letters are available online, so we can see what John Beale was sending Hartlib in the 1650s and what research he was undertaking. Is there sparkling cider still for tea? Hartlib may have died in poverty, but he started something very important, a loosely knit forum where people could freely discuss all manner of matters by letter, free from the constrictions of Oxford, Cambridge, Church and court. Democratic and radical, it had a powerful intellectual following. A format and energy which was channeled into the Royal Society.

Photographic memory

But who was Revd Beale? The great recorder of cider matters was born in Yarkhill, 9 miles east of Hereford. John Beale (1608–83) was the son of Thomas Beale (1575–1620), a Middle Temple lawyer and gentleman farmer with keen interests in orchards and cider which he shared with the Scudamores. In a letter from 1656 to Hartlib, John Beale says that his father was: 'the instrument to the plantation of our two chiefe orchards of Rotherwas, where hee was Sir Roger Bodenhams right hand, And at Yarkhill where hee planted for himselfe. There I said, my father is reported the first founder of the Redstrakt must [i.e. the Redstreak apple]. I have it so from ancient neighbours.' John Beale also inherited other orchards planted by his great-grandfather who died in 1566. So the Beales were an ancient orcharding family. Redstreak was theirs. Rotherwas is just outside Hereford.

What is also interesting is that John Beale, through his mother's family, the Pyes of Much Dewchurch, was connected by marriage to two very influential families in Somerset: the Spekes of Dillington House near Ilminster and the Phelips of Montacute House. Both estates made cider.

John Beale was obviously a bright lad, learning Ovid's *Metamorphoses* by heart before going to Eton and then King's College, Cambridge. He had a photographic memory and would scan-read books in bookshops rather than buying them. Beale stayed at Cambridge to teach and continued his own studies into hermetic and mechanical philosophy, experimenting with telescopes, thermometers and sundials. He spent three years on the Continent (1636–8), acting as tutor to his cousin Robert Pye and George Speke. In Paris they visited Lord Scudamore. On their return Edward Phelips of Montacute gave Beale the living of Sock Dennis, near Ilchester, a useful sinecure. Cash in the back pocket. Tithes. Good grazing, orchards and cheese. It was here at Montacute House that he observed and recorded the bottling of cider.

During the Civil War, Beale steered a dangerous middle path between his father's family, who were staunchly Royalist, and his wife's family, who were Parliamentarians from Shrewsbury. It was only after 1650 that he began writing about cider and orchards. These cider-making letters from the 1650s are most interesting. John Beale was one of Hartlib's keenest correspondents. Hartlib said of Beale: 'There is not the like man in the whole island.'

John Beale used Hartlib as a sounding board and communicated through him with Ralph Austen in Oxford. They had a good rapport, though Beale doubted that Oxford cider made from Pearmains and dessert fruit was anything like as good as *their* Redstreak. No surprise, then, that two of Beale's letters to Hartlib were published as a small treatise in book form: *Herefordshire Orchards, a Pattern for all England* (1656). Beale encourages everybody to plant orchards for the greater good of the country, as if it was a national duty.

Westminster cider

Another learned cider maker who corresponded with Hartlib was Dr Richard Busby (1606–95), headmaster of Westminster School who taught Classics to such brilliant boys as John Dryden the playwright, John Locke the philosopher, Christopher Wren and Robert Hooke of Hooke's Law. No doubt Busby set them texts to translate from the *Georgics* about orchards, pruning and grafting. Classics were vital for underpinning the cider revolution, giving credence and respectability to their endeavours. In 1653 Hartlib notes that: 'Mr Busby schoolmaster of Westminster is a great maker of Cider.' This was a top-notch cider school.

Another Westminster schoolmaster, Thomas Vincent (1645–56), got very drunk one night. He 'drunke boyled Cider which he professed was as delicate a drink as ever he had drunke in his life'. It just so happens that Lord Scudamore's London house in Petty France was only 300 yards from Westminster School. In June 1663 Scudamore sent Busby ten dozen bottles of cider from Holme Lacy.

More than 350 years later, in 2017, excavations at the Great Kitchen, Westminster School, revealed the remains of one such bottle with **B** for Busby stamped on it (as was the custom in those days).

Fig. 7.5 'The 'Busby' cider bottle recently excavated at Westminster School', London, *c.*1665

The neck, alas, is broken, but the sturdy onion-shaped bottle is otherwise complete and a fine relic, easily capable of surviving a secondary fermentation. Cider was on the national curriculum.

After Dr Busby's death an 'old Cyder press' was listed as belonging to him in the Cellar or Buttery, as well as 'a deal Glasse case, 4 stillings' (stands to rest barrels on) and '3 grosse of Quarts and Pints glasse bottles').

Dancing in the cup

The Beale–Hartlib letters highlight the care with which cider makers chose their fruit and pressed it, how they kept their cider cellar and bottled cider. This was the mid-1650s and predates the sparkling wine boys and champagne buffs in London by at least six years.

Here Beale is also waxing lyrical about high-tannin wild crab apples and 'peare' trees in Herefordshire which cannot be eaten, even by pigs, yet they make the best 'wine'.

Hereford Jan. 18. 1656.

> Crabs & wild peares, such as growe in our wildest & barren clefts & on hills doe make the richest, the strongest, the most pleasant & the most lasting Wines, that England yet yields or is ever like to yield as I conceive.
>
> Soe soone as I can write it, I will prove it, that the wilde apple & the peare which chokes the biter (as smartly as Aqva fortis) which neither man nor beaste pig nor fowle will bite, yields strong Rhenish, Backrac, yea pleasant Canary, at your choice sugard of itselfe or as rough as the fiercest greeke Wine, opening or binding, holding the age of one, two, three or more yeares.

Beale is comparing their best high-tannin cider to foreign wine. His philosophy is clear to see: a social contract and economic improvement suiting both

Royalist and Puritan. He blows the trumpet for cider and the common man. Like Ralph Austen, Beale is a very practical, hands-on sort of bloke. Common ground at last.

In May 1656 Beale makes the point that, because the River Wye is not navigable for barges on account of rapids, they cannot easily export Herefordshire cider. Overproduction of cider becomes a very real problem to farm labourers when temptation beckons.

'By defect of transportation Our Store of Cyder is become a snare to many, who turn God's blessing into wantoness and drunkeness.'

In the same breath Beale also points out that: 'The credit of Cyder being of few late years much advanced in the estimation of our best Gentry, who have sought out the right method of ripening and hoarding the choicest fruits, and of finding the right season of drawing it and some also of *bottling it* . . .' Just as Lord Scudamore did in the 1630s. But no one yet understood the process of fermentation turning apple juice into cider, as this 1656 quote from Sir Cheney Culpeper makes clear: 'There all Philosophers are silent which yet is nothing also but the mystery of fermentation which doth specifie and multiply etc. This is the Mystery of all mysteries.'

But did the cider *sparkle*? It was not the word they used then, but there is written evidence. In this letter from 1657, Beale doesn't mince his words. He does *not* like *flat* beer or *flat* cider. Rowdy neighbours are a problem.

16 February 1657

> We will rather drinke pure water, than the water of rottenes, as we call all drinke that does not *mantle vigorously* . . .

Mantling cider was therefore *de rigueur*, forming a vigorous head or froth, or, as the French would say, *mousse*. Flat beer and flat cider were out. Royalist youth wanted to put a bit of sparkle back into life under Puritan rule. As if drinking was for them an anti-establishment activity. A bit like walking was for Wordsworth and Coleridge. John Beale then continues his letter:

'Our Cider, if it bee brisky, will dance in the cup some good while after it is powred out: Although I disswade from all Morning draughts, yet I have brought it into fashion amongst many, They will not drinke cider, if it be no soe busy, as thoroughly to wash their eyes whilst they drinke it.'

To *mantle* is an archaic English term and predates anything in France or even London to do with sparkling wine. *Mantling cider* is where it is at in 1656–7. *Sparkling wine* comes later. The technique of bottling is, however, just the same. *Mantle* is an interesting word, from Latin, Old French and medieval English. As a noun it means a cloak, a covering, the foam that covers the surface of liquor. So it is a new phenomenon. 'To mantle' can also mean 'to flower', or 'to smile like a drink'. Spot on.

As a verb 'to mantle' means to form a head, to cream, which is just what happens. So combine that with the words 'busy', 'brisk' when 'powred out' as well as 'wash the eyes whilst they drink it' and you have an accurate description of what is going on. Sparkling is tame by comparison and does not imply great movement or energy. Diamonds sparkle but they are static. A visual trick. 'To mantle' implies movement and a deep-seated energy. A raft of very small bubbles 'that rises to the surface and forms a cloak'.

This is the earliest written evidence of the bottled sparkling technique, which was pioneered by Lord Scudamore in the 1630s then taken up by Ralph Austen and John Beale in the 1650s. They discovered it and wrote about it. They both added sugar. A technique which for business reasons in France was later christened *méthode champenoise*.

If it looks like a duck, walks like a duck and quacks, it is almost certainly a duck. If it *dances in the cup and wets the eyes* and *mantles vigorously* it must be a secondary fermentation in a bottle: that is good enough for me. Ruinart, the first French champagne house, was not founded till 1729. A full seventy-three years later.

The royal touch

These cider makers were more cutting edge than they realized. This was the holy grail, something alchemists could only dream of. Transforming flat cider into a marvel that *mantled*. It was not a chance, accidental finding. Without strong bottles it could not be done. These cider makers deserve far more recognition and praise worldwide.

Perhaps the ancient and illustrious wine world could awake from its slumbers?

Beale's letters take some deciphering and decoding. There is a sense of adventure and pioneering spirit about them. That is what I admire, for they did not have the language and knowledge of chemistry to explain things as we do today, but they noted down everything.

Beale was always experimenting with different techniques, such as leaving the pomace to rest for twenty-four hours before '*pressing the liquor out*', thus making the cider '*more rich & more pleasant*', a common practice and used in keeving. Beale then comments on '*boording the fruit in a heape*' but not for '*a whole weeke*'. A middle path before the fruit rots. Beale also mentions many varieties and he knows how to blend them. Sometimes he uses the term *harshness:* he means tannin and acidity. That makes perfect sense, helping to preserve the cider for several years. Their instincts were spot on. But Beale fully admits that most farmers in the past were very careless with their fruit and made abominable cider . . .

When Charles I visited Hereford after the disastrous Battle of Naseby in June 1645, they had a right royal cider tasting. And from that day onwards they had annual cider competition to commemorate the visit and the best cider was judged for its '*gust & wholesomenesse*'.

The word 'gust' can mean 'a sudden violent rush as with wind' or 'to taste with relish and enjoyment', from where we get the word gusto. Both fit in this case. So the cider which was once fit only for 'clownes and day labourers' was indeed fit for kings and princes. Three hundred and seventy-five years later the Museum of Cider in Hereford now has an annual cider competition in similar vein which is fitting.

Black Swan and Falcon

It was not just the aristocracy and gentry using bottles for improving and keeping their cider and giving it a bit of sparkle. Even rough old publicans got in on the act. Here Beale in May 1658 is comparing the price of cider in Hereford taverns and observing how landlords make a quick buck by bottling cider up in quart stone bottles for a few weeks:

May 31. 1658

> Nowe, for thiese two laste Monethes, cider is sold at the two chiefe Inns, the Black Swan & the Falcon, . . . They buy cider for 30s the hogshead, & sell it for 6d the Wine quarte, our hogsheads conteining 70 gallons of statute-measure, & a gallon of statute measure being about the proportion of 6 of our Wine-quarts, as I am informed.
>
> The only difference this, That at thiese Innes & houses of resort, They drawe it off in bottles some weekes before they drinke it; The bottles beeing stone botttles of a quart measure, & layd up in coole cellers.

John Beale has no great praise for their methods of perking up cider or for its quality, yet finds they can greatly increase the value of cider in a few short weeks. A miracle indeed.

> If this their botteling may justly quadruple the value of our Common cider in soe short a time,Then you may infer howe much the best cider of the longest date, may excell their 6 penny cider.
>
> And one single observation is soe newe amongst us, & yet soe powerfull,That it hath turnd our comon cider into rich Wine, & hath brought a greate gaine to them, That have the wit to reguard it.

My own hypothesis is that the cider boys in Hereford, the rough and ready publicans, had hit upon a gambit that worked well for them in 1658. With warm weather a bit of fizz would have been perfectly natural with some residual yeast stored in a stone bottle for a few weeks. Sugar may have been added but this is not mentioned. But if they were making a lot of money they would have kept that secret. Sounds like *pét nat* or *malolactic fermentation*. By early summer a quick secondary fermentation was perfectly possible, giving fizz and sparkle.

News must have eventually spread down the Thames to London and by 1662, four years later, this technique was used by London vintners to perk up their flat wines. So it appears that London wine boys had taken a leaf out of the cider manual. If they could quadruple their takings in the pubs of Hereford with stone bottles who knows what the London vintners could do. Lord Scudamore, Ralph Austen and John Beale deserve the real credit. They at least used strong glass bottles.

The Royal Society

By the time of the Restoration in 1660 natural philosophers were very thirsty indeed. Theatres reopened and comedy was all the rage. All sorts of ideas were bubbling up, as if the cork had suddenly been taken off the intellectual bottle. Puritans were old hat.

The Royal Society was founded in November 1660. It had emerged out of Gresham's College (where Sir Kenelm carried out his experiments), two secret societies, the 'Invisible College' and the Experimental Philosophy Club in Oxford, and the Hartlib Circle. Early members included Revd John Beale, the chemist Robert Boyle, Christopher Wren and Sir Paul Neile, a cider maker and astronomer. He certainly saw the stars.

The Royal Society proved very influential, not just in scientific matters, but in recording cider history. Cider was a safe topic. It eschewed politics and religion and cultivated common ground. What better place to air one's opinions about cider and its philosophy? Of the first twenty papers read to the Society at least five were on cider making and bottling cider. One paper, *On the Adulterations of Wine*, was by Walter Charleton, a Somerset doctor from Shepton Mallet. Vinters were not always held in high esteem.

John Evelyn, the diarist, was a key member of the Royal Society. His family made gunpowder, of which vast quantities were used during the Civil War. But Evelyn wielded a quill rather than a sword and 'devoted most of his time to gardening, fruit and tree culture'. He recorded the Great Fire of London and the Plague. Evelyn also had a rather low opinion of vintners, abhoring 'the Sophistications, Transformations, Transmutations, Adulterations, Bastardizings, Brewings, Trickings and Compassings' that they practised in their temples, i.e. their vaults. And of their wine – 'Let them drink freely that will; Give me good Cider.'

Evelyn praises John Beale, 'the most excellently learned Vicar of Yeovil', in helping him with *Aphorisms on Cider* which is where key information about bottling and sparkling cider is to be found.

A Walnut of Sugar

The Royal Society was in effect an upmarket cider club. The first meeting was held at Gresham's College. By 1662 Beale was openly talking about the addition of sugar to bottled cider and even gives quantities. This follows on very neatly from what Ralph Austen said in 1657. The stage is now firmly set in London.

On 10 December 1662 Beale's *Aphorisms on Cider* was read to the Royal Society. Beale puts his cards onto the table.

In Aphorism 25 he clearly states that: '*Bottleing* is the next improver and proper for *cider*; some put two or three *raisins* into every *bottle* which is to seek aide from the vine; Here in *Somersetshire* I have seen as much as a *Walnut* of *Sugar*, not without cause, used for this country *Cider*.'

This was the crunch. Beale provided first-hand, first-rate scientific evidence and gave the game away. A 'Walnut of Sugar' defined the method very precisely. I suspect he was talking about cider made at Montacute House owned by the Phelips family, to whom he was related. Montacute is only a few miles from Yeovil.

In October 2008 I was asked to give a talk at the Royal Society on sparkling cider and to reveal the information I had found lurking in their archives. When I walked through the white double-columned porch in Carlton House Terrace a few months earlier to do my research in the library, it felt as if I was entering hallowed ground. I was looking for irrefutable evidence to back up Beale's observation about '*mantling*' which he had sent to Hartlib.

As it was a Royal Society lecture I decided to do some 'scientific' experiments. So before I went up to London I put three walnuts into three egg cups. I then measured out the equivalent volume of sugar into each of the egg cups and weighed them: 18–20 gm. I then took a photograph of a 1651 wine bottle with a long neck and string lip at the top and drew an equivalent sized bottle on graph paper and counted the squares. Good old geometry, weights, measures and volumes. Euclid, Pythagoras and Archimedes, would I hope, have approved. Size: 75 cl just like today. Fortuitous.

I then did something slightly more modern: I e-mailed a sparkling wine producer down in Cornwall and asked them how much sugar they put in their wine bottles when they primed them for secondary fermentation. Back came the e-mail from Sam Lindo of Camel Valley Wines: 'We put 24g/l of sugar into the base wine which equates to 18g/bottle, 3½ teaspoons which I think is indeed a walnut.'

Bingo. The Beauty of Science. In the Royal Society's elegant library overlooking The Mall and St James's Park I then leafed through the Society's copy of *Sylva* by John Evelyn as well as original handwritten letters sent by Beale, Sir Paul Neile and Captain Silas Taylor. And there was much more in Beale's *Aphorisms on Cider*:

> In a good cellar it improves in Hogsheads the second year; in bottles and sandy cellars keeps the records of late revolutions and old Majoralities [i.e. it keeps a very long time].

> When cider is setl'd, and altogether, or almost clarified, then to make it spriteful and winy, it should be drawn into well cork'd and well bound bottles and kept some time in sand or water; the longer the better if the kind be good. And cider if being preserved to due age, bottl'd and kept in cool places, conservatories, and refrigerating springs, it does almost by time turn to aqua vitae; the bottles smoak at opening . . .

All very commendable and convincing, except in one small detail. Cider does not become aqua vitae, but it does become something else. A drink which drinks *brisky*, '*dances in the cup*' and '*wets the eyebrows*'. In fact, with the addition of sugar it would be a bit stronger, maybe 1–2 per cent more. It is a *nappy wine* – having 'a head and foaming'. Mantling again.

Pot Gun cider

As with all science it is vital to get independent verification of one's results, a bit like putting together a legal case and presenting key evidence. So I was reassured when I saw the findings presented to the Royal Society on 8 July 1663 by another of its members, a cider maker called Sir Paul Neile. Something that Dr Roger French, the distinguished medical historian for the Wellcome Trust mentions in his excellent tome *The History and Virtues of Cyder* published in 1982. Here he talks about the 'independent tradition of English cyder' in relation to wine as well as bottling.

Neile (1613–86) went to Cambridge and became an astronomer. The young Christopher Wren used Neile's 35-foot telescope to study Saturn's rings. Back down to earth Paul Neile was very scientific. In his discourse on cider, he advocated using a *nutmeg* of sugar in the bottle rather than a *walnut*. Very precise. Just like Beale, he also talks about cider '*fretting within the bottle*' ... '*a fine quickness*' that '*will mantle and sparkle in the glass*'.

It is all in the language. The word 'fret' is interesting as it means 'an agitation of the mind, a vexation, a querulous utterance a gust, a great squall'. Just imagine a great squall of cider. A year later, in 1664, the word 'fret' enters the lexicon as a 'secondary fermentation in liquors'. Just what we need. So the cider makers are not just changing the way in which cider and wine is made but also our language. To fret or not to fret ...?

But Sir Paul Neile also uses the phrase 'Pot Gun Cider' where cider '*will ferment so much that ... when you open the bottles it will fly about the house ...'* There speaks first-hand experience. Dangerous stuff Pot Gun cider. It also shows what happened if they added too much sugar. Trial and error. Sir Paul Neile builds a very strong case.

As for the timing of bottling, he is very particular and if the cider does not *fret* or *mantle* Neile's simple remedy is: 'To open bottles after about a week, that have not started on their way and put into each bottle, a little piece of white sugar, about the bigness of a Nutmeg, and this will set it into a little fermentation, and give it that briskness which otherwise it would have wanted.'

Playing on the safe side. A nutmeg rather than a walnut of sugar: 3 nutmegs = *c.*20 gm. Neile also understands the need for quality control and organized sampling: 'I commonly set the bottles in the order they were filled and so we need not open all to see the condition of the cider; but trying one at each end and one in the middle will serve the turn.' Very modern and very scientific. This is, I suspect, the earliest known case of 'sampling' and 'batch' testing, a science unto itself and a skill now determined by computers and algorithms. Sir Paul, well ahead of his times, knew what he was doing.

Most importantly Sir Paul adds crucial evidence to our case about sparkling *wine*. And this is very, very important historically. Sir Paul advocates these same methods of bottling cider for helping French wines. He was a wine connoisseur and liked vins d'Hermitage from the Rhône, Graves wine from Bordeaux on the left bank of the Garonne, as well as Verdea, a fine white wine from Greece grown from vines planted by the Venetians.

Sir Paul advises that bottling in the manner of cider '*may doe good to French Wines also*'. A crucial step forward. The first time that bottling with the addition of sugar has been properly articulated for French wine in a measured way. This is dynamite. *Wine buffs take serious note!* The pioneering techniques with bottling cider had gone from Hereford via Oxford and Somerset to London where cider makers passed on their knowledge to vintners and wine importers either wittingly or unwittingly. Then this technique filtered down to the Channel ports then across to France, where they eventually claimed the technique as their own. History is fluid. Liquid assets. *Entente cordiale*.

But did bottled cider travel well? Sir Paul recites another very illuminating anecdote about 'A Gentleman of Herefordshire this last autumn who was sending cider to London from Gloucester and had 'not enough casks'. So he used a large vat and seven or eight hampers of bottled cider with the clearest cider:

> the barque in which his cider came had a tedious passage [around Land's End]; that is it was at least seven weeks before it came to London and in that time most of the cider in the cask had wrought so much that it was much harder [drier] than it would have been.

> But the other cider, which was in the bottles, and escaped the breaking, that is by accident, had less of the lees in it then other bottles had, or was not so hard stopped, but was found to be excellent good, beyond any cider that I had tasted out of Herefordshire.

In other words the cider in the bottles had improved beyond all measure during a long, tedious sea journey, probably with a malolactic secondary fermentation. The long sea voyage may have helped as well. I rest my case.

Christopher Merret

We shouldn't forget Dr Christopher Merret (1615–95), the glass and wine expert from Gresham's College. Merret was from Winchcombe, northeast of Cheltenham. An Oxford man and a bit of a polymath, not simply a physician, he studied birds, butterflies, fossils, metallurgy and Cornish tin mines. He also translated an important Italian book on glass making, *The Art of Glass* by Antonio Neri (1612), with some of Merret's own observations. Sadly, there is little mention of bottles except to say that through the glass 'you may see their fermentations, separations and whatever other changes nature in time worketh in any liquors'.

Merret's paper on 'sparkling' wine was read to the Royal Society on 17 December 1662, only a week after Revd John Beale's and it is fascinating but somewhat vague in the weights and measures department. Not much in the way of technique either . . .

> Our Wine-Coopers of later times, use vast quantitys of Sugar and Molossus, to all Sorts of wines, to make them drink brisk and Sparkling, and to give them Spirit, as also to mend their bad tastes, all which Raisins, and Cute and Stum performe.

Molossus is molasses. *Cute* is a wine (which is wine 'boyled to the consumption of halfe') to a Butt of wine, i.e. reduced by 50 per cent to a form of syrupy concentrate. *Stum* is wine revived by new or secondary fermentation, resulting from the admixture of must or grape juice, i.e. yeast. There is also an earlier quote from the same document about revamping old wine:

> A little stum put to wine decaid [sic] makes it ferment afresh

Almost poetic . . . I am grateful to Tom Stevenson for digging up these nuggets back in 1998 and highlighting the role of Christopher Merret. André Simon had also stumbled across them in 1948. The real nub of this wine quote is that London vintners had used sugar and/or molasses to correct faulty wines and make a quick buck, just like the Hereford publicans with cider at the Black Swan and Falcon. Crucially in this quote from Merret there is *no* mention of wine bottles in any shape or form, *no* mention of corks, *no* mention of laying down the wine, *no* mention of wax or pack thread, string lips or cellars filled with running water, or even sand. Certainly no mention of strong English bottle glass. Or mantling. No hint of storage at all or maturation. Or which wines were used. Were they even French? We know not . . .

Speedy vanishing nittiness

But this new, fine-drinking, brisk, quick-fretting cider – what did it taste like? Did it actually sparkle or mantle? And if so, how much? Luckily for us sparkling cider was described in great detail in a letter read out to the Royal Society in 1663 by Captain Silas Taylor:

> I have tasted of it, three years old, very pleasant, though dangerously strong. The colour of it, when fine, is of sparkling yellow, like Canary, of a good full body and oyly: the taste of it like the flavour or perfume of excellent peaches, very grateful to the palate and stomach.

Wonderful stuff. A fine accolade from a Parliamentarian captain of horse. Silas Taylor could be talking about his own cider or even Lord Scudamore's Redstreak. And here is the crucial bit. If it was three years old in 1663 it must have been made in 1660 or possibly earlier, which predates Merret's accounts. We know it sparkles, we know what it tastes like. A true connoisseur and a composer to boot.

Taylor's job during the Commonwealth period was to sequester Royalist lands in Herefordshire. Indeed, he was so enamoured with Scudamore's cider that he was accused of going soft on the Royalists. He even delayed taking their estates off them, which suited Royalists down to the ground.

Born in Harley, near Much Wenlock, Silas Taylor was a Shropshire lad and went to Oxford, New Inn Hall, very close to Austen's cider factory. During the Civil War he served in Herefordshire, Gloucestershire and Monmouthshire. Taylor was a cider maker and had an estate at Litley Court, east of Hereford towards Hampton Bishop on the Wye, only a few miles upstream from Scudamores.They were neighbours. Silas knew all the little quirks of the trade, including bottling.

Luckily for us, he wrote down his cider-making knowledge in a long letter to the Royal Society, dated 14 July 1663. I have seen the original letter in spidery handwriting: impressive stuff. Taylor talks about the smell of apples as they ripen, how to gather them in baskets and leave them in a heap on straw till they are just right, which will 'give you such a fragrancy as is desired or endeavoured'.

Taylor then gives advice on how to bottle cider and the great care needed to get the timing right. Again, cutting-edge stuff:

> This makes it drink, quick and lively; it comes into the glass not pale or troubled, but bright yellow, with a speedy vanishing nittiness (as the vinters call it) which evaporates with a Sparkling and whizzing noise.

Sounds familiar? He gives a refreshingly honest and accurate description of what he is trying to achieve. No mere accident. Perfection itself. You can see the sparkling cider, you can taste it, you can hear it bubbling away, you can actually smell it. The Oz Clarke of his day.

Even at a distance of more than 350 years you can experience it with all your senses. Silas Taylor had a playful intellect. A true Shropshire lad, sword, quill and cider press.

As a composer his words have a certain musicality to them. Later on, Taylor was based in Harwich as keeper of naval stores and was an Admiralty spy for Pepys, reporting on the latest intelligence from the Low Countries.

'Speedy Vanishing *Nittiness* which evaporates with a *Sparkling* and *whizzing* noise.' Bravo, Silas Taylor. You hit the spot. To mantle or not to mantle. That is the sparkling English question. Again I rest my case, cider makers had indeed pioneered and perfected this new fangled bottling technique.

Méthode champenoise

But what is the French story? What is going on in the Champagne region where champagne is almost a state religion? Who invented the champagne process over there? Or did it mysteriously invent itself? What is the French version of events? In the Valley of the Marne the *méthode champenoise* story is, predictably, very different. The Marne is a lazy river and meanders slowly towards Paris. Very useful if you ship wine by barge downstream to French kings, diplomats and the bourgeoisie . . .

So what better place to start with than Hautvillers, the 'high village' above the river. The abbey has 16 hectares of vines and a large wine cellar specially cut into the chalk hillside with capacity for 500 barrels. During the late seventeenth century the cellar master was Dom Pierre Pérignon. Épernay is just down the road from where Dom Pérignon's story bubbles up from time to time.

The most fashionable wines in seventeenth century were those from Aÿ, a *cru* in the Marne Valley that became the designation for all *vins de Rivière*. Coincidentally, Aÿ is twinned with Newton Abbot in Devon, an ancient cider town. Wine bars and cider bars twinned.

The classification of vineyards is a complex business. The delicate wines from Aÿ were the drinks of French kings Leo X, Charles V, Frances I, and Henry VIII of England. Each had their own respective wine houses in or near Aÿ to obtain their own special supplies. Lord Scudamore would certainly have drunk these wines in London and when he was Ambassador in Paris. John Beale also mentions *Vin d'Aye*.

Having experimented with bottling cider at Holme Lacy, Lord Scudamore was well aware of the problems with French bottle glass which was just not strong enough to take a secondary fermentation. Also there was another more serious problem, the Franco-Spanish War (1635–9), which rampaged over the Champagne region and severely disrupted wine production.

Dom Pierre Pérignon (1638–1715) was born the son of a clerk in the town of the Sainte-Menehould on the road to Verdun. His father owned several vineyards so he grew up with vines. Pierre studied at a Jesuit college and as a Benedictine followed a strict path of prayer, study and manual labour. His escape was wine making. He entered Hautvillers Abbey in 1668, six years after the cider papers were read at the Royal Society, so he was less than a year old when Lord Scudamore was sending bottled cider to London in 1639.

For many champagne enthusiasts Dom Pérignon is a patron saint, but distentangling myth from reality is not easy. Sadly, Dom Pérignon wrote nothing down that has survived. The French Revolution may have seen to that, just as the Fire of London in September 1666 may have destroyed London vintners' records.

What is known about Dom Pérignon comes from two sources: Canon Jean Godinot (1661–1749), the son of a tanner in Reims who in 1718 wrote a treatise on wine making, *Manière de cultiver la vigne et de faire le Vin en Champagne*, and Dom Pérignon's apprentice and successor Frère Pierre who in 1724 put together a dissertation, *Traité de la culture des vignes de Champagne*. From these two works it is possible to deduce what improvements Dom Pérignon made to wine making. Revolutionary in its day, people thought he was being too meticulous. But his rules are now fundamental to the champagne industry. That alone makes him a saint.

Canon Godinot is explicit:

> One must only harvest on days when there is plenty of dew; and in hot years after a little rain. This dew gives the grapes an exterior bloom, that is called 'azur', and inside a coolness, which means that they do not heat up too easily, and that the wine is not coloured. The harvest should be begun half an hour after sunrise; and if there are no clouds and it becomes hot at around nine or ten o'clock then harvesting should stop. If the sky is overcast the harvest can continue throughout the day, because the grapes will keep well in the cool. In wet years, take care

> not to put any spoilt grapes in the baskets; and in all weathers be sure to remove bunches that are rotten, damaged, or completely dry; but never take the grapes off the bunches.

Godinot also specifies that grapes should be cleanly cut with a small, curved knife, with as little of the stem as possible, to avoid the 'taste of wood' in the wine. He also wrote: 'scarcely fifty years have passed since they [the wine growers and producers of champagne] started to work out how to make their "grey/middling wines" – which are practically white', using the Pinot Noir grape. Today the balance is nearer 45 per cent Pinot Noir, 33 per cent Pinot Meunier and 22 per cent Chardonnay.

Dom Pérignon was not fond of white grapes because of their tendency to re-ferment. He also advised the aggressive pruning of vines so that they grow no higher than 3 feet and thus produce a smaller crop. The tendency for white wines to re-ferment, or, more accurately, to finish off their primary fermentation, was a very real problem if you wanted to bottle wine. Winters were cold and in the late spring they would take off again and give bubbles. Dom Pérignon avoided bubbles. This is crucial. He hated them. They were trouble . . . They broke bottles and wasted wine.

In 1718 Godinot states that Dom Pérignon produced a wine that *moussed* (*mantled* or even *sparkled*) about twenty years earlier from when he was writing, i.e. back in 1698. This may have been a good thing or it may have highlighted a problem. This may have been *pét nat.* No sugar is ever mentioned in the abbey's records.

With French wine *mousse* describes 'the foam that forms after champagne has been poured out of the bottle. Light and airy in texture. "Soft-mousse" – not overly fizzy, but "harsh-mousse" is excessively fizzy.' Just so. Frère Pierre, a man who knew Dom Pérignon and the cellar intimately, does not mention bubbles or sparkling at all. Revealing? Serious French wine historians have also dismissed the idea that Dom Pérignon invented champagne as we know it today, but he did improve many things in the vineyard and winemaking department. And for that we are all very grateful.

Verre Anglais

Then there is bottling, which is fundamental to the whole story. You need strong green bottles, plus a good cork held in place. At that point the French had neither. Dom Pérignon blocked the neck of the bottle with a wooden plug, surrounded with oiled hemp. As a precaution the wooden stopper was fastened to the bottle with hemp string and then the neck of the bottle was dipped into molten wax. The hemp string was a forerunner of the wire attachment in use today. A document from Champagne country, written in 1718, three years after Dom Pérignon's death, recommends the use of cork. The French were often fighting the Spanish, who were running rampant in the Low Countries, but the

English had access to cork from Portugal. Thank you, Portugal.

In France, as in Elizabethan England, bottles were used to take wine from barrel to table, often with a wicker covering to protect them. If bottles exploded, they could be lethal. The French wooden peg or dowel was called a *broquelet*, wrapped in flax tow and greased with tallow. It was not until the 1690s that glass makers of the Argonne, forests near the Belgium border, imitating those in England, managed to make a thick, dark glass that enabled the transportation of Champagne's wines in bottles to any destination. This was the very region from which the Huguenot glass makers had fled 100 years earlier. Also there were restrictions on the export of wine in bottles. The monks bartered wine for bottles. Clever monks.

As for *tirage*, the addition of sugar and yeast to kick-start the second fermentation, there is a fanciful story flying around that on his death bed Dom Pérignon whispered his secret to Frère Pierre.

It was, however, Dom Grossard who started the rumour that Dom Pérignon had invented champagne. In a letter dated 25 October 1821, to M. d'Herbes, Deputy Mayor of Aÿ, he wrote, 'As you know, Sir, it was the celebrated Dom Pérignon . . . who found the secret of making sparkling and non-sparkling white wine, and how to remove the sediment from the bottles.'

At the time of the letter, Grossard was a village priest but he had never met Dom Pérignon, who died more than a century earlier. Even François Bonal, the great

Champagne historian, acknowledged Grossard's claims to be 'unfounded and even manifestly erroneous'.

What Dom Pérignon did, apart from excavating out a much larger cellar from the chalk, was to introduce the art of close pruning. This improved quality but required regular harvesting in the coolest hours of the morning, often stopping by nine. As well as harvesting every few days to select the ripest, healthiest grapes, he used smaller baskets to avoid crushing grapes. Rotten grapes were discarded. Very labour intensive. He also built press houses in various villages to reduce the distance the grapes had to be transported. These were often beam presses like the large cider presses. He made white wine from black grapes, the first known example, and started blending before pressing. It all came down to taste, palate, intuition and wine wisdom. Attention to detail and close observation, just as they were doing with cider apples in Herefordshire in the 1650s. Same principles. Sparkling cider philosophy and sparkling wine philosophy were running on parallel tracks. A shared story? I like to think so, but English mantling cider came first.

So champagne was not invented by a single individual. The evolution of *méthode champenoise* was one of prolonged trial and error with many broken bottles along the way. Obviously much of the early *méthode* was pioneered by the dedicated cider makers in England in Herefordshire, Oxford, Somerset and London. Word eventually got back to France on the grapevine about this novel sparkling technique and they went to town with it. Quite literally. It was only in the late nineteenth century that the

French perfected the technique with *dégorgement* to get rid of the old yeast cells. This required the technique of *remuage*, or riddling, by storing the bottles in specially made wooden A frames called *pupitres*, or riddling racks, where the bottles lie sideways. They are turned by hand every day then slope at increasingly steeper angles and then upside down, so that the dead yeast cells are in the neck and can be extracted easily. What is fascinating is that back in 1676 in *Vinetum Britannicum* John Worlidge not only advocated bottling cider with sugar and using corks but he advised on the way in which the bottles should be stored:

> Therefore is the laying of bottles sideways to be commended, not only for preserving the Corks moist but so that the Air that remains in the bottle is on the side of the bottle where it can neither expire nor can new be admitted, the liquor being against the Cork, which not so easily passeth through the Cork as the Air.

He then goes on to say something very interesting indeed:

> Some place their bottles on a frame with their noses downwards for that end. Which is not to be so well approved of, by reason that if there be any least settling in the Bottle you are sure to have it in the first glass.

These wooden frames, later called pupitres, are a key part of the champagne process and let the sediment settle in the neck of the bottle. In other words, the dead yeast cells, which is what the champagne method succeeds in removing so adeptly by freezing the neck of the bottle in ice. So back in 1676 the English had a very good notion of what the problems were. They even had ice houses . . .

Perhaps the best way to view this forgotten 'champagne' history is to make it a joint venture. A shared vision of sparkling perfection. The evolution of sparkling cider on one side of the Channel in the seventeenth century and the evolution of sparkling wine as a major industry on the other side of the Channel from the eighteenth century onwards.

The truth is that when Lord Scudamore started his cider-making experiments back in 1632, the genie was out of the bottle. Or to be more accurate firmly kept *in* the bottle. So without the advances of the English cider makers, the French wine makers in the Champagne region were a little scuppered.

Back in August 2008, just to try and clarify matters, I wrote to the Champagne Bureau UK in London, and asked whether Dom Pérignon did in fact invent champagne. The reply I received from the Bureau was very diplomatic. 'Dom Pérignon is credited with inventing champagne, although some authorities dispute this and maintain that it was invented earlier and probably in England.' Christopher Merret is mentioned. A foot in each camp. Very wise. Obviously back in 2008 Dom Pérignon was still in the French driving seat but the English story was not dismissed out of hand. Effervescence was the key. Very diplomatic.

In March 2021 I re-contacted the Champagne Bureau and Françoise Peretti was charming and very helpful. By now the story had shifted slightly and it was all about process.

Dear James,

You are asking who invented Champagne? To which I reply Champagne is not an invention. It is the sum of terroir, practices and know-how going back several centuries. Effervescence was neither discovered nor invented. Observed in the 17th century, the *champenois* have mastered the natural process produced by yeast that transform grape sugars into alcohol and carbon dioxide by fermentation. Over the centuries they gradually built an unparalleled expertise in this field.

Hope this helps.

Best wishes,

Françoise

As they say on their current website: 'The French distinguish between "effervescence" and "*pétillant*" (sparkling), which suggests the fizzing bubbles at the surface, or "mousse", that suggests images of a creamier, more stable substance.' Which is to my mind 'mantling'. The very word John Beale used in 1656. *Voilà*! Which is exactly what Inspector Poirot would say . . . Today Dom Pérignon is no longer prominent in the official French story. It is all about the intricacies of making champagne which is how it should be. It is big business. The French champagne industry has 33,843 hectares of vineyards, 340 champagne houses and employs 30,000 people, plus 120,000 seasonal workers. They certainly know what they are doing. The UK is their best export market.

Another very interesting angle is that in January 2019 Pierre-Emmanuel Taittinger, of Taittinger Champagne, gave an interview to *Le Figaro*, which was reported in the *Independent*, implying that champagne had come about accidentally. 'The English left these cheap, still white wines on the docks in London and the wines got cold so they started undergoing a second fermentation. Like all great mistakes, it led to a great invention.'

I would beg to differ slightly. It was no mistake at all. No mere accident. Far from it. Maybe it is high time that both England and France fully appreciated the subtle qualities of English sparkling cider and the fascinating history that lies behind this revolutionary technique pioneered from 1632 onwards using strong dark green bottles to make sparkling or mantling cider, and in the case of Sir Paule Neile sparkling wine as well. These English cidermakers have all the answers French and English winemakers have been looking for. The records of the Royal Society, Ralph Austen's book and John Beale's letters to Samuel Hartlib bear this out. Inspector Poirot would look at all the evidence and weigh it up very carefully. He does not rely on mere supposition or wishful thinking. An accident? I think not *mes amis* ...

The real heroes to my own mind are not just the English cider makers but the Huguenot glass makers who were forced to flee France in the first place. It is a shared story, and as if to reinforce this longterm association, Pierre-Emmanuel Taittinger has planted 40 hectares of land with vines in a joint venture near Faversham in Kent, which will soon be producing top quality sparkling wine. By a curious quirk of fate Selling Court Farm is only about five miles from Brogdale orchard, the successor to Henry VIII's historic mother orchard at Teynham. Another curious quirk of fate is that all the dark green, heavy-duty champagne bottles used today to make sparkling bottle-fermented cider and sparkling wine in England are now made in France. *Vive la différence*!

8

GEORGIAN CIDER

When visiting Herefordshire I always stay at Sink Green Farm on the banks of the Wye between Hereford and Holme Lacy. Lord Scudamore must have passed by many times on his horse. It is a very old site. After a fine farm breakfast I brave the traffic. All roads lead to Hereford. There below the bridge, the River Wye quietly meanders its way through town. To the right a glimpse of the sturdy, no-nonsense cathedral home to the famous Chained Library and last resting place of John Philips the cider poet. On the left, close to Sainsbury's and a Travelodge, is Bertram Bulmer's dream – the old cider headquarters – now the Museum of Cider with its old 'champagne' cider cellar down below.

The reason I have come here is to see their fine collection of Georgian cider glasses and to do more research into the 1763 cider taxation and cider riots. I also managed to have a good chat to the director, Elizabeth Pimblett. As well as running the museum, she organizes the annual cider judging competition, the paperwork for the King Offa distillery and many other cider events and displays. Elizabeth is a real live wire. A very elegant ambassador, just what cider needs these day. This is the cider embassy. Elizabeth was brought up in Hereford, her father was a local GP and they had an orchard outside town. She then read English Literature at Exeter, and did a post grad in museum gallery studies at St Andrews, specializing in archaeological collections. She has worked for the National Trust and Herefordshire County Museum Services. An ideal background for understanding the past.

As we stroll through the museum we have a long, rambling conversation about cider history: Lord Scudamore, John Evelyn, John Beale, Silas Taylor, Ralph Austen and Thomas Andrew Knight. All the usual suspects, some of whose wonderful portraits hang on the walls. The museum is home to all manner of old cider presses, even a large French beam press, as well as stone cider mills, barrels, old prints, cider labels and a new exhibition, 'Women and the Art of Cider', which Elizabeth curated. It is Tuesday and the archivist Sally Mansell is hard at work upstairs trying to recatalogue a vast collection of documents and cider ephemera. The Archive of Cider Pomology. This is the only cider museum in the country and it is a gem.

The collection of Georgian cider glasses, about forty in all, is displayed on a large table. They are Elizabeth's pride and joy: she loves them dearly. In fact, seeing them persuaded her to take the job. 'I came here in September 2016. It will soon be five years. Yes – I have enjoyed it. What sold it to me was coming round the corner and seeing that case of eighteenth-century elegant, refined cider glasses. One of the glasses has "No Excise" etched onto it. Always loved those and I thought I'd love to be able to work with those.' 'No Excise' commemorates the opposition to the infamous increase of cider taxation in 1763, a *cause célèbre* close to the hearts of all cider drinkers from top to bottom. These Georgian cider glasses, though not as tall as the Scudamore flute, were the height of elegance and more practical.

Elizabeth is ideally suited as director; she has a fresh mind, lots of energy and is prepared to look at cider in new ways. Her enthusiasm is infectious. We continue talking about cider glasses in front of the main display case. 'The large goblet in the middle with a coat of arms

Fig. 8.1 'No Excise Georgian cider glass', 1763

with fruiting apples on it was owned by the second son of the Duke of Rutland, George Manners. One glass has a barley sheaf and fruiting vines and apples. Difficult to date. Everything is an educated guess. These were very elegant, expensive glasses for the dining table and give a clue as to how important top-quality cider was.'

Elizabeth then explains the origins of the museum and how Bertram Bulmer started collecting exhibits from 1960s onwards, particularly the cider glasses. Auction houses would tip him off when they came up for sale. The Cider Museum Trust was set up in 1973 and young people did a survey of all the old horse-powered cider mills in Herefordshire. Rural cider engineering.

The museum finally opened in 1982. The old boardroom is oak-panelled and a shrine to the Bulmer family. Lots of portraits, boardroom table and chairs, the odd typewriter. Oral history recordings. Quite a small room for such a large firm. One expects Hercule Poirot or Miss Marple to come out of the woodwork. Many of the cider advertisements are from the 1930s. Very colourful they are, too, far more interesting than the modern ones. They had a bit of class about them. Quite a few had ladies pouring bottled cider into glasses, plus picnic hampers, the odd Daimler or Rolls-Royce. That sort of thing.

One of the key scientists was Dr Herbert Durham, a friend of Fred Bulmer's, from King's College, Cambridge who became director of research in 1905. Fred Bulmer (1865-1941) was Bertram's father and had turned down the plum job of tutoring the King of Siam's children to join his brother HP (Henry Percy) Bulmer in setting up his cider business. A wise move.

Dr Durham was crucial to their enterprise. He helped pioneer modern cider-making processes by isolating a wild yeast to create the first pure cider yeast culture, which ensured that fermentations were consistent. In 1906 they started making champagne-style cider called 'Cider De Luxe'. In 1916 its name changed to Pomagne, i.e. 'Champagne Cider'.

'Guaranteed Naturally Sparkling and made by the same process as Champagne – HP Bulmer. Delicious to the last drop.' With a fine-looking lady on the front holding a champagne glass. Pomagne was made like that till 1975 (when there was a court case). Then it was bulk-fermented, which was not quite the same thing. The champagne cellars beneath the museum are impressive. Miles of bottles. Acres of green glass. I can see that Elizabeth Pimblett really loves her job.

When she was 18 she was working for the National Trust and very quickly realized that people wanted to know the personal stories behind places and objects. The small day-to-day, down-to-earth details that make the place come alive. The same is true with the Cider Museum. Every object has a story and every cider family has its tales. It is the people who fascinate her most: 'We have had so many movers and shakers who have influenced the world of cider. That alone justifies the museum. Cider history is very rich indeed.' Two of Elizabeth's favourites are Lord Scudamore, who lived at Holme Lacy, and Thomas Andrew Knight. 'An amazing chap; delving into what we now call plant genetics.'

Elizabeth is particularly interested in women who have helped cider, not least by illustrating the early *Pomonas* – large illustrated books describing apples and pears. Eliza Matthews and Frances Knight provided thirty pictures for Knight's 1811 *Pomona Herefordiensis*. Some of these illustrations can be seen in the museum. The ephemeral beauty of cider apples. A real treasure trove. Still remarkably bright and fresh.

Sixty years later, in the 1870s, two equally talented young artists, Edith Bull and Alice B. Ellis, helped illustrate the Woolhope Club edition of *The Herefordshire Pomona*. Edith's father Dr Henry Graves Bull was a local doctor and past president of the Woolhope Club. Robert Hogg, a Scottish pomologist and botanist, provided notes for the 441 illustrations. It took seven years. One problem was that the ladies Edith and Alice could only illustrate a certain number of apples or pears each autumn. A labour of love and the illustrations are a joy to behold. Only 600 copies were ever produced, which are now now very valuable. The Woolhope Club was founded in 1851 and is still going strong.

This is a living museum and the events Elizabeth helps organize give a real sense of cider community, which is just what Bertram Bulmer would have wanted. Collecting, collating, storing and then displaying everything in an accessible way is what museums are all about. Bertram's daughter Gillian Bulmer is still a key member and trustee of the Cider Museum and very knowledgeable about orchards and cider. As to cider philosophy, Elizabeth Pimblett is pragmatic: 'The philosophy of cider and perry varies hugely. Today it is all about quality of fruit and not being afraid to experiment. Loads of people try ground-breaking stuff, while others will do what they have always done and be very successful. I have generally found cider makers to be quite inclusive and welcoming to me. Mainly because there is a Slow Food movement and people are more connected and interested in the orchards. When that breaks down and becomes an industrial process the quality changes. It is all about respect for apples and the process.'

As Elizabeth says, 'I didn't realize how wonderful the cider community was. Learning about cider has been a complete privilege.' Her first cat, which was rescued from a pub, was called Stowford because they were drinking Stowford Press Cider from Westons at the time . . .

Elizabeth continues: 'Our principal role in the museum is reminding people that cider can be an excellent drink with an excellent pedigree and what they may have experienced aged fifteen is only a tiny proportion of the whole story. It has a noble lineage and a very down-to-earth approach. Cider can be something to all people. The museum has a key role in UK. But people have very different levels of understanding. Some have no idea AT ALL how cider gets into their glass.'

Vinetum Britannicum

No such problems in the seventeenth and eighteenth centuries. Everyone understood exactly how cider was made. Orchards and cider apples were revered. It was heady stuff. And so before diving headlong into Georgian cider and the shenanigans of cider tax and cider riots, it is worth considering three writers whose works forged cider's popularity: the agriculturalist John Worlidge from Petersfield in Hampshire, the Sussex farmer Richard Haines and the cider poet John Philips from Hereford and Oxford. They all expounded the virtues and techniques of making cider, bottling it as well as making cider brandy. All three distilled their cider experiences into print. Words of encouragement which helped the next generation of Georgian cider makers. Even Daniel Defoe got in on the act.

John Worlidge's cider bible, *Vinetum Britannicum – A treatise of cider*, is still consulted today. It is a cider DIY guide dedicated to Elias Ashmole of the Ashmolean Museum in Oxford. Worlidge's family had been in Petersfield for several generations and lived in Worcester House in the middle of town. The second edition includes a forty-two-page discourse on bees, which illustrates beehives with glass fronts to 'observe better' how the bees behave and perform. Very advanced indeed.

Vinetum Britannicum was published with a licence obtained from Roger L'Estrange, an MP whose role was to be the Thought Police. His job was to unearth dissenting books and pamphlets. To root out nonconformist and seditious literature. L'Estrange kept a wary eye on printers and was known as the 'Bloodhound of the Press'. Luckily Worlidge and cider were above board.

Worlidge was *very* enthusiastic about promoting

cider and in his dedication to Elias Ashmole he wrote: 'this wine [cider] being that which incites some to speak too much, will I hope beg my excuse and speak for itself. It being one of the best and most advantageous pieces of improvement of our country farms yet known . . .'

Fig. 8.2 'The Ingenio', A robust scratter for crushing cider apples, no need of a horse', 1678

Note the use of the word wine. Worlidge also invented the ingenio, a man-powered cider apple-crushing machine which speeded things up and didn't require a horse. One or two ingenios are still in operation today. As always, Worlidge was partisan and encouraged apple growing: 'Cider hath been improved to perfection, as many ingenious and worthy persons can testifie.'

'I hope every Englishman, or native of this Isle will

conclude with me, that our British Fruit yields us the best Beverages; and of these Fruits, the Apple is best, which is here called Cider.'

Worlidge describes *seider* and *sicera* in ancient times. Then mentions Androclyes writing to Alexander the Great to lay off the wine as it makes him very rude. He praises Lord Scudamore for the purity of his cider and his ability to keep cider for many years. The most interesting parts of Worlidge's book are the fine details, for instance, cleansing barrels, which he describes almost poetically.

First, 'To make you cask pleasant to receive so delicate a guest as your choicest Cider, you may scent it as the Vintners do for their wines. Thus take Brimstone, four ounces, of Burn'd Allum one ounce and Aqua Vitae two ounces; melt these together in an earthenware pan over hot coals, then dip therein a piece of new canvas and instantly sprinkle thereon, the powders of Nutmeg, Cloves, Coriander and Aniseeds. This canvas set on fire and let it burn in the bunghole, so as the fume maybe received into the vessel.'

Purify. Purify. This is good advice. Same today. Keeping wooden barrels clean was a nightmare and still is. Then he discusses glass bottles versus stoneware bottles with salt glazes, some of which were imported from Germany.

'Glass bottles are preferred to stone bottles because the stone bottles are apt to leak, and are rough in the mouth, that they are not so easily corked.'

As always, Worlidge is practical. He prefers corks to glass

stopples which have to be ground with emery paper, first rough then smooth. He then goes into more detail about bottling and, of course, adding sugar where needed. Further on Worlidge gives good advice on corks: 'Great care is to be had in choosing good corks, much good liquor being absolutely spoiled through the only defect of the cork; If corks are steeped in scalding water a while before you use them, they will comply better with the mouth of the bottle than if forc'd in dry.'

Worlidge also has a stab at distilling: 'After fermentation, the spirits become volatile which is the reason that most Liquors are most pleasant after fermentation and then also they are capable of yielding their Spirits by Distillation. For then you may, after due fermentation extract spirits, vulgarly called *Brandy* in great plenty, and very excellent quick and burning . . .' But that's another story . . .

Cyder Royal

Most people know 'cider brandy' by its French name, Calvados or Calva, which is specific to the Calvados region in Normandy. In cafés in the cattle market in Rennes, Calva is automatically added to your coffee to increase the tempo of the bidding. You have to put your hand over the cup to refuse. Not often done. Cider brandy in this country also has a long history but was often called aqua vitae, the 'water of life', cider's answer to whisky or cognac. Then there was Cyder Royal using cider brandy with cider 50:50, more like a port, and was 'invented' by Richard Haines in 1684. Haines was an ardent pamphleteer.

Distilling cider or wine was nothing new. Alchemists, wizards, monks, priors, chemists and natural philosophers like Roger Bacon had been at it for centuries, often with the help of the Arabic texts translated from the Greek. But none ever turned it into a business. In France, Gilles de Gouberville, Seigneur of Le Mesnil-au-Val in northern Normandy, distilled spirit from pot stills in 1553. On 28 March 1553, in his journal, he mentioned the practice of distilling cider to make apple brandy. He is revered as the patron saint of Calvados.

Another reference appears in Somerset in 1560.The will of Robert Gibbes, last prior at Montacute Priory who leaves to his nephew '. . . one *lymbeck*, one *stillatorie* and xx dosens of glasse'.

'To my lad James Kitto: one fether bed, one bowlster, one pillow . . . one stillatorie, one brasen potte to make *aqua vitae* in, my little ambling mare with bridle gurses and sturrops with all such books as I have of Physicke and Surgery.'

Robert Gibbes was a doctor and distilled herbal concoctions as well as aqua vitae in the priory. He must have taken the distilling equipment with him. Fast-forward 100 years and there is even a book on the subject.

John French (1616–57), a physician, wrote *The Art of Distillation* in 1651. He went to New Inn Hall, Oxford, and applied chemistry and alchemy to medicine. His mentor, Lord Saye and Sele of Broughton Castle, recommended him as physician for General Fairfax in the Parliamentarian army. In *The Art of Distillation* French dubbed distilling the 'true naturall philosophy' which, he suggested, ought to replace 'that empty naturall philosophy which is read in the Universities'. High spirits indeed. French's book draws heavily upon the work of Hieronymus Brunschwig, a botanist and chemist noted for his methods of treating gunshot wounds, and for his early work the *Liber de arte distillandi de simplicibus* (1500) on distillation techniques. The book is both practical and finely illustrated. French distils anything and everything, even rotten apples: apple grappa:

WATER OUT OF ROTTEN APPLES IS MADE THUS

> Take as many rotten apples as you please. Bruise and distill them either in a common cold still or gourd glasses in Balneum. This water is of greater use in fevers and hot distempers than the common distilled waters of any cold vegetables. It is very good in any hot distemper of the veins and sharpness of the urine. It is very good in the inflammations of the eyes.

What more does a man need? There is a peculiar and wonderful essence that black rotten apples give off, very delicate and fragrant which belies the sight. French Calvados from the 1960s still has that beguiling pear-drop aroma.

Fig. 8.3 'Primitive but effective distilling apparatus', 1651

HOW TO MAKE AQUA VITAE AND SPIRIT OF WINE OUT OF WINE

Take of what wine you please. Put it into a copper still, two parts of three being empty. Distill it with a worm until no more spirit comes off. Then this spirit will serve for the making of any spirits out of vegetables, but if you would have it stronger, distill it again and half will remain behind as an insipid phlegm. And if you would have it yet stronger, distill it again, for every distillation will leave behind one moity of phlegm or thereabouts. So shall you have a most pure and strong spirit of wine. A Hot Still.

Wine can of course in those days can mean cider as well. John French died near Boulogne serving with Cromwell's army in 1657.

At the same time a Sussex farmer from Sullington called Richard Haines (1633–85) was experimenting with distilling cider to make cider brandy and then adding it back to a hogshead of ordinary cider to make his own fortified cider like sherry, port or Madeira wine. This was *Cyder Royal*.

Haines entered into partnership with Henry Goring of Wappingthorn in Steyning. His patent was granted in 1684. Haines understood the process very well: 'take eager, very hard or sowre Cider (for that yields much more spirits) twelve gallons; distil it as other spirits are distill'd, in a Copper Body and Head, and a refrigeratory Worm running through a Cask of cold Water, under whose Beak a receiver is placed. From which with a gentle fire, draw off two gallons of Cider brandy, or spirits . . .'

Richard Haines's treatise and tract was printed and then advertised in London in a broad sheet available for 6*d*.: 'Aphorisms upon the new way of improving cyder, or making cyder-royal lately discovered for the good of those kingdoms and nations that are beholden to others, and pay dear for wine by Richard Haines.'

In other words, he was fed up with paying through the nose for imported fortified wines and as a farmer he could see the economic potential for having a home-grown distilling industry in Sussex. Why not distil your own cider brandy? It was either that or get entangled with smugglers.

Richard Haines died in 1685 aged 52. Alas, his ideas of

Fig. 8.4 'John Philips, The Cider Poet' *c.*1700

a large internal and overseas trade in cider brandy and Cyder Royal did not emerge. Cheap Dutch gin and smuggled French brandy saw to that.

The cyder poet

Imagine on a winter's evening sitting beside a log fire with a glass or two of cider brandy reading cider poetry. In 1708 John Philips wrote forty pages of epic blank verse, *Cyder a Poem*. An ode to cider. His style is sandwiched somewhere between Virgil and Milton, a rustic view of rural life. The classical world creeps into every page. The timing was just right. Philips captured the spirit of the times and received high praise.

John Philips (1676–1709) was born in Bampton, Oxfordshire. The family estate was at Withington, east of Hereford. Prime orchard country. His father was the vicar, so he knew cider inside out. Philips studied natural history and botany at Christ Church, Oxford. Many fellow students were descendants of Royalist cider makers.

But Philips was never in good health, suffering from asthma. Smoking did not help. At Brasenose, he knew the principal's daughter, Mary Meare, for whom he did 'conceive a secret passion'. An 'agreeable brunette without anything very striking except a full black piercing eye'. She disdained him and in solace he began writing poetry. Philips stayed ten years at Oxford without taking a degree and published the *Splendid Shilling*: a Miltonic 'ode to poverty' commended in *The Tatler*. Queen Anne was on the throne.

In *Cyder*, Philips describes the art of making different shaped glass bottles and compares single variety ciders maturing into 'sparkling perfection'.

The Moyle, And tastful Pippin, in a Moon's short Year,
Acquire compleat Perfection: Now they smoke
Transparent, sparkling in each Drop,

Stirom or Styre apple makes the strongest cider which keeps for up ten years. But there is a word of warning about drinking too much cider when bottles fly and people are injured with broken glass:

Now horrid Frays
Commence, the brimming Glasses now are hurl'd
With dire Intent; Bottles with Bottles clash
In rude Encounter, round their Temples fly
The sharp-edg'd Fragments, down their batter'd Cheeks
Mixt Gore, and Cyder flow:

About *Cyder* the botanist Philip Miller said, 'there were many books written on the same subject in prose, which do not contain so much truth as that poem'.

Sadly, in 1709, only a year after the poem was published, John Philips died of TB, aged 33. He wanted the name of cider to go round the world. And now it has. Philips is buried in the North aisle of Hereford Cathedral. His friend Simon Harcourt also erected a marble monument in Poets' Corner in Westminster Abbey. He is flanked by Chaucer and Drayton. Fame indeed for a cyder poet.

Daniel Defoe

After Queen Anne died in 1714 the Hanoverians took over, and the Georgian age began with a flourish. In London, the German beer Brunswick Mum was flavour of the month but it did not dent sales of Georgian cider. Cider boomed and was shipped into London from South Devon, Sussex and the Channel Islands. This was cider's heyday. A whole century. Four Georges in a row. Royalty was back in business. Cider was on dinner tables and in salons. It was very fashionable to drink cider out of crystal glass engraved with apples and pears with twisted stems, white spirals and blossoms. Amazingly, a few glasses still survive.

At the other end of the social scale there were late-night Georgian drinking dens called 'cider palaces'. J. M. W. Turner grew up in Maiden Lane in Covent Garden, reeking of flowers, opera-goers, actors, pickpockets, wig makers, barbers and ladies of the night. In the basement of Turner's house was William Wootten's Cider Cellar, a drinking den described in 1750 as a 'midnight concert room', a cellar which had cider-fuelled operas all of its own: *Rake's Progress*, *Redstreak does it Again*, *M'Luds Little Pippin. A Nice Pear*.

Another talented London artist, James Ward (1769-1859), was born poor and like Turner knew all about cider. Ward's father was a cider and fruit merchant who fell on hard times due to drink, leaving his mother destitute. Young James was put to work filling bottles of cider, then washing bottles rather than being sent to school. For him this was a blessing in disguise for it allowed his imagination to wander. James had time to sketch the lively scenes around him and develop his artistic skills. His brother-in-law was the artist George Morland who specialized in painting animals. Cider and art intertwined

Another imaginative person was Daniel Defoe (1660–1731). As a young man, Defoe was captured at the Battle of Sedgemoor in 1685, at Westonzoyland in Somerset. The heart of cider country. This was during the Monmouth Rebellion and Defoe was on the losing side. He only just escaped with a pardon and his life. It cost a fair packet.

The Duke of Monmouth landed in Lyme Regis and gathered an army of tradesmen, weavers, carpenters, farm workers and cider drinkers, armed with scythes, hayknives and pitchforks. They did not like James II. Rumour has it that on the eve of the battle they were imbibing too much cider in Bridgwater. Easily done. Many lost their lives either in the battle or in the Bloody Assizes afterwards.

If Judge Jeffreys had been drinking Somerset cider it would have dissolved his kidney stones and he might have been more lenient on the rebels. 'Don't waste the Court's time by pleading ye innocence.' Either way, cider drinkers suffered and many were transported to Barbados, where in the north of the island they still speak English with a Somerset accent. Maybe all cider makers have a rebellious, nonconformist streak hidden beneath the surface.

Defoe certainly had a taste for cider and commented on it favourably several times in his writings and the vast quantities of cider flowing into London. Here is a letter from his West Country travels in the 1720s:

> There is one article in the produce of Devonshire, which makes good what I have written before, that every county contributes something towards the supply of London; and this is, the cyder which I have mentioned already, and which takes up the south part of the county, between Topsham and Axminster, where they have so vast a quantity of fruit, and so much cyder made, that sometimes they have sent ten, or twenty thousand hogsheads of it in a year to London, and at a very reasonable rate too.

In 1724 he visited Herefordshire:

> As for hops, they plant abundance indeed all over this county, and they are very good. And as for cider, here it was that several times for 20 miles together, we could get no beer or ale in their public houses, only cider; and that so very good, so fine and so cheap, that we never found fault with the exchange; great quantities of this cider are sent to London, even by land carriage, though so very remote, which is an evidence for the goodness of it, beyond contradiction.

In Somerset, he writes about cheddar which 'without all dispute, is the best cheese that England affords, if not, that the whole world affords'. Cider and cheddar, yet again.

Royal Wilding

Back in 2007 I went to Faversham in Kent and spent a day at Brogdale Horticultural Trust and did some research in their library. On the bottom shelf at the back I found a wonderful book from 1729: Batty Langley's *POMONA or the Fruit Garden Illustrated* to which was added *A Curious Account of the Most Valuable CYDER-FRUITS of Devonshire*. This addendum was sent by Hugh Stafford of Upton Pyne in Devonshire, just north of Exeter. A fine apple growing area with red soil.

Batty Langley's book was excellent and had many illustrations. Apples were taken very seriously. Langley was a well-known garden designer when landscaping was all the rage. Alexander Pope lived down by the river with his grotto. Even Voltaire was knocking around in Wandsworth. 'Judge a man by his questions rather than his answers.' Maybe he should have judged a man by his apple trees and cider . . . Voltaire would no doubt have supported a man's right to drink good cider and would have defended the rights of small artisan cider makers.

Batty Langley also kick-started the 'gothick' revival, but what really caught my eye was Hugh Stafford's letter, a fascinating example of how cider apples can encompass people's minds. If only Scudamore's Redstreak had such attentions. In my own copy of the letter the paper has been attacked by a bookworm, which makes it look like one of Batty Langley's eccentric garden designs. It is an extraordinary epistle to a single variety of apple and shows the care that Georgian cider makers took with their experiments.

A Curious ACCOUNT of the most Valuable
CYDER- FRUITS of DEVONSHIRE

To Mr Batty Langley at *Twickenham date Nov 10th 1727*

Sir

Since you have seen the *Royal Wilding Apple* itself which is so very much celebrated (and so deservedly) in our County . . . The single and only Tree from which the apple was first propagated, is a very tall, fair and stout one, which I believe about twenty Feet high. It stands in a very little Quillet of Gardening, adjoining to the Road that leadeth

> from *Exeter* to *Oakhampton* in the parish of St *Thomas*, . . . A walk of a mile from *Exeter* will furnish any one, who hath such a Curiosity with a Sight of it.
> IT appears to be a proper *Wilding* . . . and hath in all probability, stood there more than seventy years. [i.e. since 1658]
>
> IT is a very *constant* and *plentiful* bearer every other Year, and then produceth apples to make one hogshead of Cyder which contains 64 wine gallons;
>
> And having tasted it, and found the Juices, not only in a most perfect *Soundness* and *Quickness,* but such likewise as seemed to promise both the *Body* and *Roughness* and *Flavour* that wise Cyder Drinkers in *Devon* now begin to desire.
>
> Mr *WOOLCOMBE* was not a little pleased with it, and retained the name of *Wilding,* and as he thought it superior to all others . . . hence the triumphant *Royal Wilding.* The colour of the *Royal Wilding* is a bright *Yellowish*. The other qualities are a *Noble Body*, an *Excellent Bitter*, a *Delicate Roughness* and a fine *Vinous Flavour*.
>
> I HAVE only to add concerning the *Royal Wilding* that within these twelve or fourteen years, I believe more than 200,000 of the grafts have been propagated in *This* and the *Neighbouring Counties*; and if I mistake not, I heard about two Years since, that some of them were sent for from *Yorkshire;*

No mean achievement for one wild Exeter apple.

Hugh Stafford also wrote *A Treatise on Cyder* which was published posthumously in 1753 alongside a second anonymous cider treatise. But where has the *Royal Wilding* gone? A mid-season bittersweet that can make excellent single cider. A few trees apparently still exist in trial orchards . . . also known as *Cadbury* or *Pounsett* in Somerset. But you can never be sure. Rumour has it that it has been rediscovered in New Zealand, but this may be a red herring. The cider world is full of such stories. It all comes down to taste and matching DNA.

Super celestial cider

Hugh Stafford (1674–1734) was an enthusiastic cider pioneer. His country seat Pynes House, a large red-brick Queen Anne house, is still poised above the River Exe with its orchards. In his treatise on cider, Stafford speaks of his second favourite cider apple, the White Sour of the South Hams. A small yellow apple from between Dart and Teign, an early ripening apple that matures in August. 'The fruit produces a potent juice, agreeable to lovers of rough cider', i.e. high tannin. 'The White sour can be mellowed by racking.'

Competition between ciders producers was fierce, yet good-natured and amicable: between the Royal Wilding and the White Sour from the South Hams. 'Each gentleman did not contend, as is usual, that his was the best cyder; but each affirmed his own was the worst; the gentleman of the South-Hams declared in favour of the *Royal Wilding*, and the gentleman of our parts in favour of the *White-Sour*.'

But Hugh Stafford had another wilding cider up his sleeve: the apple trees grown from Redstreak pips and planted on his estate. He spent years bringing his pioneering cider project to fruition. That is how things should be done. Trial and error. Gribbles, wildings and crabs. Hugh Stafford picks up the tale again:

'The new apple was christened by one of his friends the *Super Celestial*.' But what did this *Super Celestial* from Upton Pyne taste like? Stafford indeed flattered himself one year that he had beaten the Royal Wilding. Although thinner than the Royal Wilding it was regarded as 'very excellent and admirable' . . . but was 'too *brisk* or *frisking*. . .', i.e. a sparkling cider, 'whilst the *Royal Wilding* preserved all its native virtues'. '*Brisk*' means 'effervescent', or sparkling which was perhaps seen as a fault in Devon. Maybe its tendency to break bottles played against it. *Brisk* and *frisky* – just what you want in your cider for your *hindes*? The cider quadrille. A cider fandango?

This is fascinating because it implies that the Redstreak, as well as its Devon wilding, the *Super Celestial*, had an innate tendency to sparkle. So Exeter could have had its very own sparkling cider factory on its doorstep in the 1720s. But obviously they preferred flat cider in those days. Horses for courses. Flat racing versus steeple chasing. Exeter racecourse is to the south. Some prefer a little sparkle, dancing in the glass. Some have a long finish, others are short and crisp. Take your pick.

Today, almost exactly 300 years later, someone does indeed make sparkling cider north of Exeter. Just up the valley from Pynes House, Polly Hilton from Find & Foster at Huxham Barns, picks up cider apples from the very same orchards which once belonged to Hugh Stafford. The estate is still held within the family. I went there in the autumn with Polly and the pomologist Liz Copas to help identify the remaining cider apples. Sadly, no sign of Royal Wilding. But the sparkling bottle fermented cider was excellent with a chunk of Mary Quicke's very local cheddar cheese.

Meadyate and Cockagee

The fascinating thing was that many other cider apples were mentioned in Hugh Stafford's treatise: sadly, most do not exist today which makes them even more interesting . . .

'The *Meadyate* is a very constant and plentiful bearer every other year. The juices of it have all the body and roughness of the two other cyders . . . The apple has something of the taste of Fox-welp, it is of a yellow colour, some faintly streaked in Red; it ripens very late, it is said to have its birth in the parish of Ermington near Ivybridge near Plymouth. The name derives from *meade gate*, or meadow gate.' The cider was definitely challenging:

'The *Meadyate* has this peculiar quality, that where it is but the twentieth part in making a hogshead of cyder, it will be predominant; it is therefore very useful to mix with ordinary apples which are apt to make heavy insipid cider. When the cider of which this Apple is made is received into a mug, there generally appears a *bluish dew or bloom round the sides of it*. Of this apple is made that sort of liquor, which in *Devonshire* is called *Hewbramble* or *Bramble Cyder*; alluding to its roughness which causes a sensation as if a bramble had been thrust down the throat and suddenly snatched back again.' Very high tannin. Sadly, it is no longer around. Maybe its descendants are.

The Backamore from Plympton St Mary outside Plymouth is 'beautifully streaked with dark red and has a bloom on it like a plumb. It makes an excellent strong and palatable Cyder. Frequent rackings will render it agreeable to the lovers of sweet cider. It makes a little before Christmas, an excellent tart and sweetmeats, having a touch of the Quince.'

Cowley Bridge Crab is 'dirty purple and green'. 'Old trees exceedingly fruitful. One spreading tree has been good for 70 seasons [i.e. it goes back to *c*.1650s]. Six hogsheads of cider off one tree.' Austere. Alas, it has also vanished.

Then there is Cockagee, from Ireland with a most interesting back story. As the author says, 'This fruit is of Irish extraction, the name signifying in their language *Goose-turd*. I have tasted it and (as Mr Stafford says) I find nothing extraordinary in it. Tis true it has a golden pipinary flavour. Counsellor Pyne, a gentleman who resided near Exeter and had the care of Sir William Courtenay's estates in Ireland, is said to have brought it into England. It was propagated in Somersetshire around Minehead before its name was known in Devonshire.'

What is interesting is that Sir William Courtenay's estates in Ireland were about twenty-five miles southwest of

Limerick. One possibility is that cider apples were taken from Devon in the late sixteenth century and came back over a century later. The description of cider made from Cockagee actually fits a modern Cockagee cider made by the Seed Savers of Co. Clare, which I was lucky enough to taste in Drogheda. Amazing that a description tallies up 300 years later. A light, summery cider. Just right for a hot day when out scything hay.

The origin of the name is that the apple looks a bit green and yellow, like goose shit. Another story is that a learned man was taking down notes in his diary about apple trees and there sitting under the tree was a goose amid its own shit. And when the man pointed to the tree and asked what it was called, the local wit naturally enough said, Cockagee, and so the name stuck.

Cider tax

One of the occupational hazards of producing something as eminently drinkable as cider is that sooner or later the Government comes to hear about it and starts knocking on your door. They want their cut. Over thousands of years governments have perfected the art and usually concoct complicated regimes to ensure payment. Governments needed extra finance for wars, paid for in arrears by taxing those who could least afford it. And this in addition to two other taxes that affected trade.

In 1690 a new tax on glass was introduced to help pay for yet another war against France. Tax on bottles 1*s*. a dozen for quarts and *6d.* a dozen for pints. Bottles for sparkling cider – also 1*s.* a dozen. A rise of 60 per cent. By 1696 many glass makers in Bristol, Gloucester and Newnham had gone out of business. Cider makers chose to sell it 'in Cask rather than run the hazard of putting it in bottle at so dear a price'. The Cider Bill was repealed in 1699. There were also four Navigation Acts between 1660 and 1696 which constrained the wine trade from France. This suited the cider boys no end.

Secondly, in 1703 the Treaty of Methuen, known as the 'Port Wine' treaty, with Portugal reduced the tax on Portuguese wines, i.e. port, and allowed greater trade of English woollen goods. Portugal duty free. In return, Portuguese wines imported into England would be subject to a third less duty than wines imported from France. This meant that even if we were at war with France, port and Madeira still came our way. Essential liquid assets.

Then there was cider tax, a fluid arrangement which started in 1643 when the Long Parliament decided to tax cider at the rate of 1*s.* 3*d.* a hogshead. John Pym's excise ordinance also covered tobacco, beer, ale, cider, perry, raisins, currants, figs, sugar, playing cards, thread and silk. Between 1644 and 1645 beef, mutton, veal, alum, copperas and hats were also taxed.

Bureaucracy was set up to ensure payment of taxes and collectors could resort to military assistance. This is the bare bones of a system that we still live with today: HMRC – Customs and Excise – tax inspectors, revenue men armed with staffs and gauging sticks plus cudgels and firearms if smugglers are involved.

John Pym (1584–1643), a friend of the Drake family, was MP for Tavistock. But Parliament was desperate for money, so Pym reluctantly brought in the first cider tax in July 1643. With Restoration in 1660 everything changed, but Charles II, knowing he was on to a good thing, instead of abolishing cider tax increased it from 1*s*. 3*d*. to 2*s*. 6*d*. a hogshead. Half a crown in real money; 10*s*. a tun for imported cider. That's royalty for you. Payment was confined to retailers thus excusing 'Gentlemen' cider makers from 'tedious' visits by excisemen which were considered 'inconvenient and a little demeaning'. Big estates made a lot of money selling cider, so in 1697 the cider tax was increased to 4*s*. the hogshead.

Fast-forward to the 1750s and the Seven Years' War against the French. George III is on the throne. Seventeen fifty-nine: the Year of Victories, with Wolfe cliff-climbing in Quebec, as well as battles at Minden, Masulipatam, Quiberon Bay, Pondicherry and Guadeloupe. Basically, Britain acquires a larger slice of India and Canada and scores Florida off the Spanish. The Quebecois cider makers are a bit miffed but have new customers for their cider.

Seventeen fifty-nine: the keel of HMS *Victory* is laid down and Smeaton's tower, known as the Eddystone Lighthouse, is first illuminated, which helps desperate seamen returning to Plymouth on dark and stormy nights in search of a good pint of South Hams cider. Wrecking of a different kind.

The Treaty of Paris in February 1763 brought the Seven Years' War to an end. The de facto prime minister, Lord Bute, also a resident of Twickenham, scratched his head (under his wig) and wondered how he might pay for it all. Bute elbowed aside the Duke of Newcastle and Pitt the Elder to take up the reins of power. So far so good. But then he made a fatal mistake. He not only decided to tax cider more heavily but to tax people in their own homes regarding private consumption. This was his undoing. He had crossed an invisible line.

Taxation on commercial cider had been rumbling on since 1643. Every year or two the same, almost identical documents were printed. They are small works of art: 'An Act for continuing and Granting to His Majesty, certain duties upon Malt, Mum, Cyder and Perry, for the Service of the Year 1757 Malt *6d* a bushel, Mum 10*s* a barrel, Cyder and Perry made for sale 4/- per hogshead.'

Back in 1725 the malt tax was extended to Scotland at half the English rate, 3*d*. a bushel. This nearly wrecked the fragile Union. The Scots were unused to having their beer taxed. Malt tax riots ensued. So beer was as incendiary as cider, at least north of the border. The problem was that West Country men, by drinking cider, evaded the malt tax. Nonconformist in both senses of the word. Cider was a totem, a ritual object worthy of veneration.

The debt for the French war was £146 million. Only £137 million was funded, leaving a £4.7 million shortfall. Bute wanted to knock the debt on the head quickly, so he extended taxation of cider right into the home. The head of each household was required to prepare a list of names of every family member over 8 years old, and pay 5*s*. per person per year to the exciseman, a vast sum for poor farm labourers. With a family of six, that was 42*s*. up front. It crippled them.

In default of payment, they would have to pay 4*s.* a hogshead for cider found on the premises. There was an exemption on houses that were rated below 40*s.* and made less than four hogsheads of cider, but they had to register.

Commercial cider makers had to pay an extra 4*s.* a hogshead on top of the 4*s.* but the excisemen were 'permitted to enter the millhouse, storehouse, warehouse, cellar and all other places . . . used by any person or persons whatsoever, either for making, laying or keeping of Cyder or Perry . . . to gauge and take account'. Any person avoiding payment could be fined £25 for each offence. A whacking great fine equivalent to about £2,500 today, or in those days: 3 horses, 5 cows or 250 days' work. That was for each offence. Draconian.

There was uproar in the West Country from top to bottom. At one moment outright anger, at others 'doom and gloom'. This tax was seen as a gross invasion of privacy and traditional liberties. Riots and public meetings were so outspoken and socially wide-ranging that Lord Bute resigned two months later, in April 1763, even before the Act came into force. Cider country and its vociferous inhabitants had ousted an unpopular prime minister in less than eight weeks. A record. The simple lesson to Parliament: if you mess with cider unfairly in the West Country, this is what happens . . .

Cider riots, 1763

Cider riots and political reactions to the increased duty were fascinating. It was not just a rabble. Many large landowners had vested interests in cider. It was big business. Lord Bute was not a cider drinker, so he had no idea what cider meant to the majority of people in the West Country. He may never have ventured west of Twickenham. Certainly he underestimated the depth of feeling not just about cider but the wholesale infringement of human rights and civil liberties.

Giving the excisemen the liberty of entering your property without a warrant, unannounced, at any time of day or night to search for cider, as if it was contraband, then tax it and fine the cider maker and his family was a step too far. Effigies of Bute and excisemen were hung in public places and then burned, as if for real. In Somerset the collective memory of Judge Jeffreys was still very strong. Lord Bute had touched a raw nerve. Peasants were getting their own back on a pernicious government.

Civil unrest was understandable. As well as mock executions there were speeches, demonstrations, broadsheets and broadsides, pamphlets, placards and cartoons, awash with lampoons, ditties, sarcasm and political innuendo. The press had a field day. It was a wholesale revolt against authority which went on for three years until the Act was repealed in 1766. Revolution in all but name.

But there was also an important psychological and

social dynamic at work. It was the invasion of privacy that upset people most of all. Cider drinkers were unruly at the best of times, but in Somerset there was dissident nonconformity. A sense of fair play which was deeply ingrained in their character. Cider helped to maintain a sense of independence, identity and integrity. It balanced the books, it kept the countryside stable and thirsty towns in good fettle. But if that sense of fair play was taken for granted or thwarted then all hell broke loose.

This injustice was felt at many levels. In the West Country you make cider on your own farm or small-holding. You are king of all you survey, even if it is only an acre of orchard. You are proud of your self-sufficiency and that engenders a slight stubbornness which senses injustice and values personal freedom. It also resents the intrusion of too much authority. In Somerset it also comes from living on the Levels where summer and winter grazing have their own rules, where farmers lead a more mobile, self-sufficient, hunter-gatherer lifestyle. Cider orchards anchor them to one place and keep them afloat.

Accounts of riots are interesting. They were spontaneous and widespread. The rural economy functioned on cider. It oiled the wheels of commerce. Tithes were paid in cider, rents were paid in cider, debts were paid in cider, wages were paid in cider. There was talk of a Scottish Yoke, the Scottish Boot; a man in a kilt was often seen to be tied to a stake and burned. Guy Fawkes all over again. There were 'No Cyder Act' teapots, 'No Cyder

Act' drinking glasses. Even 'No Excise on Cyder' longcase clocks.

Bute also made enemies by ousting William Pitt. Bute was an outsider imposing a 'Norman'-style yoke on 'English' yeoman. Committees were formed of landowners, meetings were held, resolutions passed, letters sent to London, MPs bombarded, pressure exerted on Parliament. There were Cyder Lords and Cyder Bishops. The knock-on effects were felt far and wide, even in America.

The cider tax was patently very unfair. Geographically it not only affected the West Country, but it also affected the price of cider much further afield as it hit the middlemen who shipped cider into London. Large landowners were incensed. People simply refused to pay the tax increase.

Politics can be very local but no less powerful. This all-pervading sense of injustice fuelled riots and protests in Cornwall, Herefordshire, Devon, Somerset, Gloucestershire and Worcestershire. Crowds even set upon the hated tax collectors. The Government proceeded to send an army to the West Country to subdue the people, but the tide of resentment was too strong. The countryside was in open rebellion. There was a religious intensity to it. Thousands marched in bereavement bearing symbols of freedom and mourning. Church bells were stilled. Effigies were hung in every town and then burned on bonfires. These mock executions had a powerful effect on the populous. They thought the cider tax wasn't fair and they were quite right.

'Liberty, Property and No Excise.'
'In Seventeen hundred and Sixty Three
Let the Cider tree from Tax be Free'

Of *Freedom* no longer, Let Englishmen boast,
Nor *Liberty* more be their favourite toast:
The *Hydra* OPPRESSION your *Charter* defies
And galls *English* Necks with the *Yoke* of Excise.

In Vain you have conquered, my brave Hearts of Oak
Your *Lawrels* your *Conquests*, are all but a *Joke*;
Let a r-s-ly PEACE serve to open your eyes
And the d-n-ble Scheme of CYDER-EXCISE.

I appeal to the Fox, or his friend JOHN A BOOT.
If tax'd thus the Juice, then how soon may the *Fruit*?
Adieu then to good *Apple-puddings* and *Pyes*
If e'er they should taste of a cursed EXCISE.

The Peace is Good. Who dare dispute the fact?
See the Fruits thereof – The CYDER TAX.

People in Exeter carried white wands symbolizing purity as opposed to the black staff of the excisemen.

At a Publick House, a figure was prepared, the lower half of which represented Jack-Boot, the upper part in a plaid bonnet with a Star. This figure was exposed to view all day. About 6 o'clock began another procession in the manner following: First a man riding an ass, and on his

> back an inscription *From the Excise and the Devil Good Lord Deliver us.* A Strip of apples in Mourning was hung around the ass's neck, and was supported by thirty or forty men, each having a white wand with an apple on top of it, also in Mourning. Next came a cart with gallows fixed to it, and the plaid figure hanging by the neck. After that came a cider hogshead, with a pall over it carried by six men in Black cloaks. On the pall was a number of Escutcheons and Inscriptions to the effect as those above mentioned. The whole was accompanied by some thousands of people, hallowing and shouting though the principal streets of the City and at night a bonfire was prepared into which they cast the figure and burnt it to ashes.
>
> *Bath Chronicle*, 19 May 1763

This feels like a forerunner of the French Revolution without the bloodshed. Cartoons appeared of a dwelling in Herefordshire being surrounded by excisemen and every door being barricaded and women emptying chamber pots from upstairs windows over the heads of the men below.

There were cartoons of the 'Roaster Excisemen' or 'Jack Boot's Exit', which portrays a man on the gallows and a fire under him fuelled by illicit alcohol. Sometimes Bute is seen wearing a kilt with tartan socks, a tam o' shanter on his head and a set of bagpipes slung over his shoulder.

The middle classes and landowners were equally incensed by the Cider Act. They were liberal-minded people and drank tea as well as cider.The message to Government was loud and clear from all angles. In the Forest of Dean the cider makers and the 'subterranean gentry' captured an exciseman and held him underground in one of their mines for over a month. There was a detailed report in the *Berrows Worcester Journal* (founded in 1690 and one of the oldest newspapers in the world) dated 21st July 1763:

'Great numbers of Excisemen took up their qualifications at our Quarter sessions this week - Some perhaps may repent of their Employ.As we are told the Subterranean gentry of the Forest of Dean are determined to take all that come within their Reach to the Regions below - One of the Bretheren of the Stick was last week catched by these Sons of Darkness in going his rounds to the Cyder Mills, and was instantaneously hurried down 2-300 feet underground, where he now takes up his Abode. The colliers it is said, use him very well, and he lives as they do; but they swear the Day of his Resurrection shall not come to pass till the Cyder act is repealed or at least till Cydermaking is over . . .'

The Cider Act was also very strict about the movement of cider. You could not move more than 6 gallons of cider at any one time from one place to another, 'even from the Mill into the cellar without a permit from the Excise officer, under penalty of forfeiting the cyder and the Cask or vessel containing it'.

In other words: 'Look over your shoulder.' Cider with Kafka? Or Cider with Big Brother? After three years of discontent in July 1766 the Cider Act was repealed.

Fig. 8.5 'No Cyder Act Teapot', Yorkshire Creamware with crabstock spout and reverse inscription, 'Apples at Liberty', *c.*1766

Commercial cider was still taxed but the smallholder was not. There was rejoicing up and down the country. Long dinners and speeches fuelled by cider, of course. Cider was a potent symbol, having a central place in Everyman's heart. Not to be tampered with. A hogshead of fine cider was more valuable than the Crown Jewels. For a short while cider had become its own political party.

Excise records

There were three unexpected legacies from the cider riots, apart from broken teeth, bruises, hangovers and doors wrenched off their hinges. The first, a tall, slender cider monument in Somerset, secondly the Stamp Act in America, which replaced the Cider Act and unwittingly lit the fuse for the American Revolution, and thirdly the excise records themselves from 1763 to 1766. Quite a legacy.

These meticulously kept excise records, when put together, give a complete cross section and breakdown of the eighteenth-century cider industry county by county. This shows the balance between small exempt cider makers, those that paid duty by hogsheads and those 'compounded', which was the amalgamation of land *and* cider tax in one large hit. *Berrow's Worcester Journal,* produced a list of *Hints for Cydermakers* which included: *Reasons against Compounding*.

The cider tax records are important historical documents that chart the scale and distribution of cider making nationwide. These were kept meticulously and the revenue balanced out against the wages of excisemen. According to their calculations there were 103,760 cider makers in England and Wales. The population was 6,736,000: one cider maker for every sixty-five people!

Excise Register 1764–5	**Exempt duty**	**Makers not compounded**	**Cider makers compounded**	**Total Cider makers**
Barnstaple	1,198	1,367	4,694	7,259
Bath	1,384	589	4,349	6,322
Bedford	4	192	48	244
Bristol	310	70	911	1,291
Bucks	5	476	126	605
Cambridge	131	764	145	1,040
Canterbury	84	537	391	1,012
Chester	–	–	–	–

Excise Register 1764–5	Exempt duty	Makers not compounded	Cider makers compounded	Total Cider makers
Cornwall	837	1,317	3,210	5,361
Coventry	–	16	1	17
Cumberland	–	–	–	–
Derby	–	5	–	5
Dorset	1,005	1,257	2,938	5,200
Durham	–	–	–	–
Essex	12	280	452	744
Exon (Exeter)	600	53	6,058	6,711
Gloucester	1,882	124	4,292	6,298
Grantham	–	–	–	–
Hants		919	208	1,752
Hereford	2,690	58	8,297	11,045
Hertford	21	533	433	987
Isle of Wight	237	229	107	573
Leeds	–	–	–	–
Lichfield	–	30	5	35
Lincoln	–	7	–	7
Liverpool	–	–	–	–
King's Lynn	16	289	17	322
Manchester	–	–	–	–
Marlborough	943	903	941	2,707
Northampton	–	92	3	95
Northumberland	–	–	–	–
Norwich	71	752	66	889
Oxon	210	920	428	1,558
Reading	34	301	341	756
Richmond	–	–	–	–
Rochester	691	1,768	1,685	4,144
Salisbury	917	931	827	2,675

Excise Register 1764–5	Exempt duty	Makers not compounded	Cider makers compounded	Total Cider makers
Salop	8	191	213	412
Sheffield	–	–	–	–
Suffolk	92	271	113	476
Surrey	34	595	273	902
Sussex	1,402	1,941	1,027	4,370
Taunton	1,618	129	6,390	8,137
Tiverton	472	153	5,443	6,068
Wales East	423	594	985	2,002
Middle	920	575	1,762	3,257
North	–	–	–	–
West	–	–	–	–
Westmoreland	–	–	–	–
Woolhampton	995	122	1,994	3,111
Worcester	1,537	581	3,165	5,283
York	–	–	–	3
London	–	–	–	–
TOTALS	21,408	20,012	62,340	103,760

What this shows is very interesting, rather like the end-of-season batting averages for county cricket. Devon comes top by a long chalk because of maritime cider, then Somerset, then Herefordshire, which has the highest number of small exempt cider makers, which is what you would expect. Herefordshire bats way above its size and still does, but has no coastline for export. It also shows how widespread cider making was and how many depended on cider income for their livelihoods. Some figures had to be 'compounded' as there was more than one excise town in counties such as Devon, Somerset, Kent and Wiltshire.

County	No. of cider makers	County	No. of cider makers
Devon	20,038	Kent	5,156
Somerset	15,750	Sussex	4,370
Hereford	11,045	Berks	3,867
Gloucester	6,298	Mid-Wales	3,257
Wilts	5,382	East Wales	2,002
Cornwall	5,361	Hants	1,752
Worcestershire	5,287	Oxon	1,558
Dorset	5,200	Cambridge	1,040

Many well-known cider counties are in the middle range. Wales has more cider makers than you might expect, tucked away on remote hill farms. Even Oxford and Cambridge feature. All those spires and deep intellectual thoughts in need of inspiration. Thirsty dons. Maybe there are still a few bottles of sparkling cider lurking in their cellars . . .

The Cider Monument

One very elegant legacy of the cider tax that can be seen today is the 140-foot-high Doric column at Curry Rivel, near Langport, with a fine view of the Somerset Levels. Designed by Capability Brown and built of the finest Portland stone. Its builder was the aptly named Philip Pear.

The slender monument, complete with internal staircase, was commissioned by William Pitt the Elder, grandson of 'Diamond' Pitt, to commemorate the landowner Sir William Pynsent (1679–1765), MP for Taunton and High Sheriff of Somerset. Pynsent, a wealthy cider maker, so admired Pitt's politics and his stance over the cider tax that he left his entire estate to Pitt, though he had never met him. Some say he persuaded Pitt to fight the issue and if Pitt won then he would give him his estate. In his will, dated 20 October 161, he observes: 'I hope he [Pitt] will like my Burton estate, where I now live, well enough to make it his country seat'. Pynsent died aged 85.

Fig. 8.6 'The Cider Monument', Burton Pynsent, Curry Rivel, Somerset, designed by Capability Brown, Portland Stone, 140 ft high, 1767

It is a fine monument and an appropriate testament to the victory of common sense over Government bureaucracy. Locally it is known as 'the Cider Monument'. What is fascinating is the nature of the political debate and how Pitt swung it his way. Pitt was out of power and in opposition, but he sensed a golden opportunity and uttered his famous statement in defence of the cider maker: 'The poorest man may in his cottage bid defiance to all the forces of the crown. It may be frail – its roof may shake – the wind may blow through it – the storm may enter – the rain may enter – but the King of England cannot enter.'

In other words 'an Englishman's home is his Castle', a phrase which Pitt used, echoing the earlier words of Richard Mulcaster, headmaster of Merchant Taylors' School, who said pretty much the same thing back in 1581.

So the background history of the Cider Monument revolves around Customs and Excise, William Pitt the Elder, Capability Brown, a local landowner as well as cider tax and erosion of civil liberties. The very bones of political philosophy entwined with cider making. As to political economy: income from Pynsent's estate was £3,000 a year and a fortune of £30,000 in cash. William Pitt did quite nicely, thank you. The Cider Monument cost £2,000.

And here is the other fascinating part of the cider story: in the 1760s the awakening of self-awareness, republicanism and self-determination was becoming an issue not just in England but in France, America and Ireland. Taxing cider became the burning political issue of the day. If the Government could not get the money it needed for the Seven Years' War from cider, it had to come from somewhere else, so they imposed Stamp Duty with the Stamp Act of 1765. This fell on the American colonies like a thunderbolt, the first time that tax in the colonies had to be paid directly to London. The reaction was pretty much the same in America as it had been in the West Country with the Cider Act.

There was widespread rioting and even looting. American colonists got a taste for it. They read the radical newspapers from England about the cider riots and developed an appetite for dissent. A secret society was formed called Sons of Liberty. Effigies of officials were burned and some excisemen even caught, tarred and feathered. Law and order broke down. The Stamp Act had lit a fuse. There were many confrontations which sowed the seeds of the American Revolution. By 1773, when the Tea Act was enforced, the scene was set for the Boston Tea Party. The rest is history.

So rioting in the South West of England about the Cider Act led indirectly to the Stamp Act, which fuelled the American resentment against London, which then led to the American colonists fighting the War of Independence. *Olé*.

But the American story has yet another twist. The debates about the rights of excisemen to enter and search your property were later set down in US law in the 4th Amendment to the US Constitution. This prohibits unreasonable searches and seizures. The legal reasoning behind this amendment derives from the debates held in the British Parliament over cider tax and warrants issued to suppress anti-Bute publications, especially those written by John Wilkes and John Entick. The 4th Amendment also sets down requirements for issuing warrants which must be issued by a judge or magistrate. The warrant must give reasonable grounds for a search and state the places to be searched and goods to be seized. A Bill of Cider Rights.

As to financial matters, Pitt's son, William Pitt the Younger, had his own battles to fight, but he drew a healthy income from the Pynsent estate, i.e. partly from the sale of cider. Later, as prime minister, he introduced income tax, which still lives with us today. So the cider debates which raged in 1763 and perceptions of personal freedom are as relevant today as they were then. The Capability Brown tower is not called the Cider Monument for nothing. It is in fact a monument to cider philosophy as well as civil liberties.

A wonderful 'No Excise on Cyder' longcase clock still keeps good time in Julian Temperley's kitchen at Burrow Hill Cider farm in Kingsbury Episcopi. The tall mahogany-veneer clock was made in 1763 by Joel Spiller in Wellington. An eight-day clock with engravings relating to cider and the opposition to the cider tax plus a few hogsheads, a sun and a parrot. NO EXCISE ON CYDER ticks away quietly in the background as we have a cup of tea. From the top of Burrow Hill you get a very good view of the Cider Monument, only 3 miles away, and the Somerset Levels.

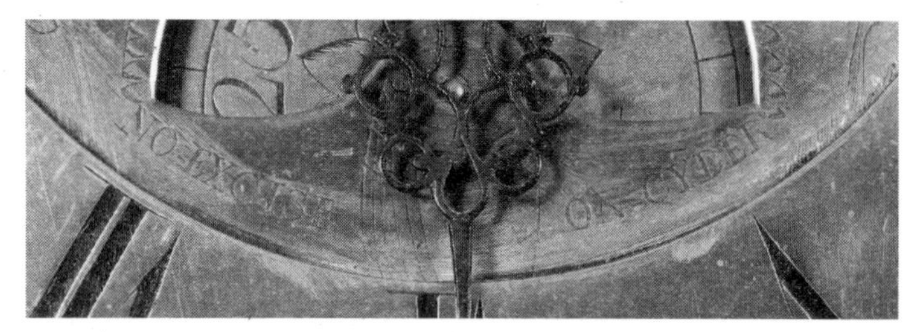

Fig. 8.7 'Detail of 'No Excise On Cyder', eight day long case clock made by Joel Spiller of Wellington, c.1763

Cider taxes were eventually abolished in 1830 by the Duke of Wellington, whose own monument can be seen in the distance above the town of Wellington where the cider tax grandfather clock came from. It seems governments only introduce cider taxes when they are in real trouble. During the First World War cider was taxed at *4d.* a gallon and lifted in 1923.

Cider tax was reintroduced in 1976 by Denis Healey. The background story, which I heard recently from someone very senior in the civil service, was that in the early 1970s they were asked to look into the idea of taxing cider again. Not just because of Europe, but closer to home. A certain well-known cider maker in Hereford had taken out an advertisement which drew attention to the fact that cider did not attract duty in that year's budget whereas beer did. They were raising 'A glass to the Chancellor'. That was a fatal bit of marketing which backfired not just on them but on everybody else in the cider world. An own goal. Even the Government of the day sat up and took notice and said to itself,'Lets' see what shall we do this year, Minister. Hmm Why not tax cider again?' And so they did just that. 'No one will notice . . .'

Luckily they had learned their lessons from 1763 and small cider makers were given exemption if they made less than 1,500 gallons. How they arrived at this magical figure I am not sure, but assuming a farm worker back in the 1960s drank a gallon of cider a day, that is 300 gallons a year with Sunday off to sober up. A farmer and four workers is five men. Well, 5 x 300 = 1,500 gallons. It has been a very useful stepping stone for artisan cider makers and has contributed to the present cider renaissance. Cider legislation has it uses. The high quality of these artisan cider makers is a monument to their skills.

Devonshire colic

Devonshire colic was a strange phenomenon seen in Devon and to a lesser degree in Somerset, Gloucester, Hereford and Cornwall. It only seemed to affect cider drinkers. The main symptom was stomach ache, hence the name colic. Its causes were debated by physicians and cider drinkers for many years.

The first written account of Devonshire colic comes from Dr William Musgrave (1655–1721) from Nettlecombe in Somerset. Musgrave's publication *De arthritide symptomatica* (1703) included the first scientific description of Devonshire colic, and was used by John Huxham (1692–1768) of Totnes who set up practice in Plymouth.

Symptoms of cider colic were observed by John Philips, the cider poet, back in 1708:

> The must of palid hue, declares the soil
> Devoid of Spirit; wretched He that quaffs
> Such wheyish Liquours; of with Colic pangs
> With pungent colic Pangs distressed he'll roar
> And toss, and turn, and curse th'unwholesome draught.

Unpleasant, yes. But the real symptoms of Devonshire colic were often far worse, as Dr Huxham outlined in 1739:

> This disease began its attack by an excessively tormenting pain in the stomach, and epigastric region, with an unequal weak pulse, and coldish sweats; the tongue in the mean-time was coated with a greenish, or brown, mucus, and the breath was most offensive. An enormous vomiting soon followed, for the most part of exceeding green bile, sometimes black, with a great quantity of phlegm excessively acid and very tough; nay the foul matter brought up was oftentimes so very acrid, that, by excoriating the throat and the oesophagus, it was tinged with blood, and created a difficulty and pain in swallowing . . . Things continuing in this state for a day or two . . .

As vomiting abated, Huxham describes muscular pains, terrible pains in the bones, weakness and subsequent paralysis that brings a fatal conclusion. Huxham thought the disease was caused by the consumption of large quantities of fresh, rough cider, and this was the accepted cause for Devonshire colic. Pretty gruesome stuff and no mistake. Not a good advert. Not like Devonshire cream teas and Dartmouth crab sandwiches. There was no real cure apart from laying off the cider.

The riddle was solved by another Devon doctor born in Modbury, in South Hams. Sir George Baker (1722–1809) had a smart London practice and counted among his patients Sir Joshua Reynolds, George III and Queen Charlotte.

What George Baker noticed was that symptoms of Devonshire colic were very similar to a disease called 'Poitou', which was often associated with wine drinking in western France. Poitou colic was described in 1616 by François Citois. In 1757 it was demonstrated in a publication by Théodore Tronchin of Geneva, to be a form of lead poisoning. There was Normandy colic as well.

In 1767 George Baker put forward his hypothesis to the College of Physicians. He observed, first, that symptoms of Devonshire colic were similar to lead poisoning. Secondly, lead was used in the cider-making process in Devon to seal joints in the circular granite cider mills, stone troughs and holes in wooden launders, acidic cider apple juice dissolving the lead to form lead acetate.

Lastly, George Baker conducted tests on cider with Dr William Saunders (1743–1817), a physician from Aberdeen and expert in chemistry, to demonstrate the presence of lead. Baker also contacted Dr Wall of Worcester who noticed that a farmer one year kept cider in a lead vat and that cider gave the same symptoms. An obvious clue which proved the point.

And here was yet another interesting twist. Cider was not the only culprit. As late as 1764, in *Secrets belonging to the Mystery of Vintners*, advice was given to vintners that they could sweeten wine with lead. So John Evelyn's detrimental comments on wine in *Sylva* were not that far wide of the mark.

Sugars of lead or lead acetate was commonly used in Roman times to help sweeten and preserve wine on long voyages. Grape juice was boiled up and reduced in lead vats, then added to the wine. Not a good thing to do. Archaeologists have looked at lead content in Roman bones and can show where the lead isotopes in the wine came from, as far afield as Eastern Europe.

But the Romans had their own cure for lead poisoning. Taking warm baths in Bath. What could be nicer. Taking the waters in Bath was very fashionable. Hospital records still exist for curing Devonshire colic and these were quoted by George Baker in his famous paper.

'Between 1762 and 1767, 285 patients with Devonshire colic were admitted to Exeter hospital. 209 were cured or improved. During the same period, 281 patients were admitted into Bath hospital, of whom 259 were cured or improved.'

Back in 1760 Dr Rice Charleton began classifying the results. They fell into six groups: 'Cured, Much improved, No better, Improper, Irregular (patients discharged before treatment was completed for misbehaviour, or at their own request), and Dead.'

Patients went to the baths once or even twice a day for long periods of time. No doubt reading the *Bath*

Chronicle. And if they felt up to it, they could dance in the Assembly Rooms which opened in 1771. In Devon there was enormous resistance to the findings in Baker's paper, but eventually it was accepted as a fine piece of inductive reasoning. Health and safety in the cider world. Medical and chemical philosophy coming to the aid of cider.

Doctors now knew the cause of Devonshire colic and remedied that quickly, but they did not know why the cure in Bath worked so well. That was only discovered in the 1970s via a very unusual source linked to NASA space research into weightlessness. A medical research scientist at Southmead hospital Bristol called Audrey Heywood, took three workers from the Avonmouth lead smelter with high levels of lead in their system. They sat in water for several hours a day and she monitored their responses. Her paper shows that workers excreted high levels of lead in their urine, which helped cure them, this was facilitated by relative weightlessness in water. Audrey Heywood then wrote another paper about the eighteenth-century treatments for lead poisoning in Bath. Her conclusion was that:

> By using modern techniques it is possible to suggest how these cures could have been achieved. Sitting in warm water up to the neck formed a vital part of the treatment in Bath, and this immersion could have contributed to the cures claimed by the Bath Hospital.

In other words a modern explanation for the success of the Bath remedies in curing Devonshire colic.

Crabbing the parson

Not as dangerous as Devonshire colic but still unpleasant is the apple-throwing custom called 'crabbing the parson' which in the nineteenth century became a raucous, anarchic event. 'Crabbing the parson', held in Worcestershire, took place on St Kenelm's Day, 17 July (St Kenelm was a young eighth-century saint who was murdered by his relatives), when the village felt that it is necessary to throw crab apples at the parson. Pelt would be a better word. Poor parson. The reason for this odd custom comes from a vignette contained in *The Rambler in Worcestershire, or, Stray Notes on Churches and Congregations*, by John Noake. Written in 1848, it says this about the parson, who got what he deserved: 'For many years, villagers at Kenelstowe in Worcestershire celebrated St Cynehelm's Day (July 17) with a village fair and the ancient custom of 'crabbing the parson' – bombarding the unfortunate cleric with a volley of crab apples.'

Thank you, Eleanor Parker – *A Clerk of Oxford*.

The parson in question was travelling to the church to give a service and, while taking shelter with a villager, felt a little peckish and secretly stole a few apple dumplings and concealed them in the sleeve of his surplice. These then fell out one by one onto a parishioner's head from the pulpit while he was conducting the service. When the service was over, the clerk let fly at the parson with some crab apples he had with him which he had collected for his horse. Hence the retribution. Wizened old schoolteachers were often called 'crabby', a bit hard and sour. Mind you, crab apples are often used as rootstocks and are often planted in orchard hedgerows to aid pollination.

Jersey cider

One island where cider was made in vast quantities was Jersey, only 14 miles from the French coast. I was invited there several times by Julia Coutanche of the Jersey Heritage Trust and I was able to make recordings with cidermakers and research their fascinating archives. Jersey has twelve parishes, two languages, an ancient Norman French legal system, a bailiff and a very long cider history. Some of the laws are archaic. Knitting was banned in church because the needles made so much noise. Men were forbidden to knit during daylight hours in the fishing and harvest season. Jerseys were useful, orchards were plentiful and the export of cider prodigious. At one point there was a law preventing the planting of more orchards, because if the island was ever besieged by the French they would need fields for wheat and barley to feed themselves and their animals.

Jersey was an important stepping stone between England and France, convenient for both sides both culturally and economically. Camden's *Britannia* (1586):

> The midland part of this isle is somewhat high and mountainous; but the valleys under these hills are finely watered with brooks; and very pleasant being planted with fruit

> trees, but apple trees especially of which they make Cyder. The villages stand thick and make in all twelve parishes, which have the advantages of many fine creeks for ships.

In 1664 in Evelyn's *Pomona*, John Newburgh from Bridport writes about Jersey cider being made from pure juice that flows naturally from pulped apple held in the 'cider cheese' whilst standing in the press before pressure is exerted:

'In Jersey they value it a crown upon a hogshead dearer then the other.' This information he takes 'from a neighbour of mine who lived in that Island which (for Apples and Cider) is one of the most famous of all belonging to his Majestie's Dominions.'

'His Majestie' in this case is Charles II. One thing that makes Jersey cider different from English cider is that they leave the pomace to rest overnight for twelve or even twenty-four hours, so that it becomes brown and oxidized before pressing. This method in Devon was known as *keeving*. This mellows out the tannins and makes the cider smoother. Similar to cider making in Normandy.

According to Newburgh, Jersey cider was held in high esteem, no doubt due to climate, location and cider apples. Historic quotes show that Jersey was part of the diocese of Winchester in 1204 when Jersey cider was a source of income for Battle Abbey. Large quantities of cider were shipped from Jersey to Winchelsea around 1270 and again to Southampton and Poole. Wool one way for knitting and cider the other.

A listing for stores for Mont Orgueil Castle in 1469 includes payments to merchants of Caen for cider, and that in 1529 two and a half pipes of cider were brought from Normandy. A pipe is about 110–120 gallons, i.e. two hogsheads. Cider was easily transported by sea. The milder maritime climate of Jersey also aided the formation of blossom in years when late frosts might have affected orchards in England. Greater sunshine led to higher sugar levels in apples and high alcohol content, which meant that cider kept better in barrels. Jersey apples were used to add quality to cider in Dorset and Devon.

The expansion of the cider trade in the seventeenth century led to the States Act of 1673 which forbade the planting of orchards except for the replacement of old ones. The upsurge in orchard planting was as a direct result of concessions granted in the reign of Charles II. Cider imports were banned from France. This then secured an effective monopoly and every available field was planted with apple trees. Corn gave way to apples. At the same time cod fisheries were starting in the Gaspé Peninsula on the south shore of the St Lawrence River in Canada. Salt cod and cider. In some parts of Canada today they still speak Jèrriais, the traditional language of Jersey people.

In Charles II's reign, 24,000 hogsheads were produced annually and a third was exported. In 1676 a decree was made giving the Channel Islands freedom to trade with England with no import taxes or harbour dues. A reward for supporting the king during the Civil War.

Cider export was big business and barrels were in great

demand. In 1801 figures from Revd F. Le Couteur give an overall production figure of 30,000 barrels of which 20,000 were consumed locally and 10,000 were exported. Ten thousand barrels is a vast amount of cider, over half a million gallons.

In 1853 142,240 gallons of cider were exported as well as 99,700 bushels of apples. In a bumper year much of this went to Devon and when the barrels ran out they simply exported the apples. Figures are always slightly suspect and the production may well have been greater than declared. What is not in dispute is that Jersey had cornered the trade in cider and apples to southern England for over 200 years. There are some wonderful stories about fishermen out at sea being inundated by clouds of blossom off the east coast of Jersey and at night being able to navigate their way home in the autumn by the rich, aromatic smell of cider making from certain farms.

As for Jersey, their pride and joy is the 'Richmond' map, commissioned by the Duke of Richmond and surveyed in 1787. Four plates were engraved in 1795. Access was very limited - with French invasion on the cards. So secret was this map that it was not publicized until 1845.

The main delight for cider makers is that the Richmond map is so accurate that every field and orchard is marked, so that you can see the extent of orcharding in each parish and thus cider production. The local land measurement in Jersey is the *vergée* (just under half an acre). The total area of orchard in 1795 is 9,916 *vergées*, equivalent to 4,402 acres. At this time the area of orchard in Dorset was 10,000 acres and Dorset is far larger than

Jersey. Some parishes in the east were over 35 per cent cider orchards.

As to yields, if you allow 5 tons to the acre of apples this yields 22,010 tons, giving approximately 3,301,500 gallons, which is 61,000 hogsheads. Revd Francis Le Couteur, founder of the Jersey Agricultural Society, gives 30,000 hogsheads, so the yield may have been only 2.5 tons to the acre, but orchards are biennial, cropping heavily every other year. The peak export was in 1836 when approximately 320,000 gallons were exported which is 6,000 hogsheads. As well as this, about 230,000 bushels of apples were exported. Jersey cider was also advertised in the *Bridport Times*, *c.*1850. Cider coming in through West Bay. Bridport was famous for rope making, net making and boat building. Even Channel Island smuggling vessels were constructed on the slips at West Bay, alongside revenue cutters.

The word 'Jersey' is often used in Somerset bittersweet cider apples: Burrow Hill Jersey, Ashton Brown Jersey, Red Jersey, Broadleaf Jersey, Early Red Jersey, Coat Jersey, Stembridge Jersey, Harry Master's Jersey, White Jersey, etc. But according to pomologist Liz Copas, these bear no relationship to any apples found on Jersey today, which are mainly bittersharps. *Jaisy* is also a Somerset dialect word for cider apple.

At the end of the nineteenth century, as cider exports tailed off, potatoes took over – *pommes de terre* – the Jersey Royal being introduced and then propagated by Hugh de la Haye around 1880. Sadly, many apple orchards were grubbed out during the German occupation of

1940–45 to grow crops for food. Starvation was only just round the corner. Apple trees were used for firewood. One pro-German Irishman, was often seen cycling between farms with a small distilling apparatus strapped to the back of his bicycle. Jersey apple poteen? Or was it cider brandy? Cider philosophy in Jersey has it own maritime flavour. They even have black Jersey apple butter, a wonderful concoction brewed up late at night with spices and stirred for days on end. When fishing out on the cod, that must have tasted glorious.

Thomas Andrew Knight

One man who carried out a vast amount of research on cider apples was the horticulturalist Thomas Andrew Knight (1759–1838), born at Wormsley Grange, northwest of Hereford. Seventeen fifty-nine was, as we have already seen, the Year of Victories. So he was a real Georgian, dying six weeks before the coronation of Queen Victoria. His father was vicar and his family made their money by running an iron foundry. They had substantial properties scattered across Herefordshire. Thomas went to Balliol College, Oxford, then lived at Elton Hall, near Ludlow, and started his experiments with fruit, vegetables and cattle breeding. Thomas Andrew Knight helped to develop many of the fruit and vegetables that we still eat today. He met Joseph Banks in 1795 and they corresponded. Banks persuaded him to share his findings and so he began sending articles on grafting and diseases of fruit trees to the Royal Society.

Fig. 8.8 'Thomas Andrew Knight, horticulturalist, botanist and cider maker' (1759–1838)

In 1797 he published *A Treatise on the Culture of the Apple and Pear, and on the Manufacture of Cider and Perry*, which was reprinted several times. One of the many things Thomas Andrew Knight introduced was controlled pollination for plant breeding. He recognized that new types were needed for orchards to replace existing varieties. He introduced the 'Grange' apple in 1792, 'Downton Pippin' apple and 'Waterloo' cherry. His plant breeding experiments extended to peas, strawberries, pears, potatoes, cabbages, plums and nectarines.

Around 1809 Knight inherited Downton Castle from his brother and used the 10,000-acre estate for experiments in plant breeding. He built an extensive greenhouse and moved into a cottage in the grounds.

As a plant physiologist Knight was fascinated by nature, orchards and cider. He was interested in what we now call 'genetic traits' and he observed the inheritance of susceptibility to disease of apple and pear trees. Very advanced work for the times. Knight had even gone so far as to determine that there were three compatibility groups for breeding.This has subsequently been confirmed in the twentieth century due to the genetics involving chromosome numbers.

Knight was alarmed by the decline in traditional varieties of fruit and considered it a matter of urgency to breed new varieties. He looked at hybridity, fecundity and heredity. Close, accurate observations. Most of these researches were published by the Royal Society and RHS.

Thomas Andrew Knight was definitely on the right genetic track. He researched inherited traits in peas by crossing various varieties. By keeping careful notes, he found many of the same results as Mendel did. If Knight had thought more holistically about the significance of his findings, rather than the simple need for improvement in plant breeding, he might well have pipped Mendel to the post, twenty-five years before Mendel was born. Maybe not enough cider was involved.

Thomas Andrew Knight observed mixed ratios in the offspring of crosses between peas with different phenotypes, or easily observable characteristics. But unlike Mendel he did not count the numerical results. Statistical analysis would have shown that there was an underlying pattern which could only be explained by the existence of a pattern or binary genetic code, which we now call genes. Even computers work on binary code, a strange language that underpins our world these days. Interestingly, Darwin had read Knight's book and it had quite an influence on his thinking. Darwin credits him in the first chapter of *The Origin of Species* for his pioneering work on plant breeding.

However, it was Thomas Andrew Knight's beautifully illustrated volume *Pomona Herefordiensis* that was the cat's whiskers. Published in 1811, it had thirty aquatint plates engraved and hand-coloured by William Hooker, Elisabeth Matthews and Frances Knight. They were works of art. Glowing images of apples pears, some familiar, some not: Redstreak, Golden Pippin, Foxwhelp, Red Must, Hagloe Crab, Loan Pearmain, Grange Apple, Orange Pippin, Downton Pippin, which he had bred himself, Woodcock, Oldfield Pear, Forest Stire, Teinton Squash Pear, Foxley Apple, Pawsan, Best Bache, Yellow Elliot, Longland Pear, Old Quining and many more.

Knight became president of the London Horticultural Society, a role he held for thirty-six years until his death in 1838. Apples and cider were at the heart of his research. But apples are so complex genetically that any attempt to unravel their genetic code would have ended in failure.

Without Knight's systematic and scientific approach it would be impossible to disentangle the inner secrets of apples and pears, secrets which are far from simple. Darwin knew full well that Thomas Andrew Knight was on to something. Even today, 200 years later, it is a complex task. Peas were a doddle by comparison.

More cider, vicar?

The Georgian age came to an end with the death of George IV in 1830. His brother William IV took over the helm and then in 1837 it was Victoria's turn. What the Victorian era saw was a shift from farmhouse to cider factory, a new business dynamic that needed hydraulic presses, sales staff, advertising and marketing. Many cities had thirsty workers and railways meant easier distribution. Vicars were quick off the mark to give advice and take up the challenge. At least three large, well-known cider companies were started by clergymen or in the drawing rooms of rectories. Orchards and tithes. God and the cider apple intertwined. They never missed an opportunity.

One rectory in Somerset stands out: Heathfield, a few miles west of Taunton between Norton Fitzwarren and Milverton. Revd Thomas Cornish (1763–1840) was rector for fifty-three years. He arrived in Heathfield in 1786 aged 23 and died there at 77. Having come from Kingston St Mary, his interest in cider making was a vital ingredient to his parish. Just to be sure, he planted his own orchard of 15 acres on the rich red loam of Taunton Vale.

In 1840 Heathfield had 136 inhabitants, tithes were commuted and glebe land was 62 acres. Revd Cornish's oak barrels in his cellar were all stamped TC – 1805. A well-known story is that during one church service his foreman ran from the rectory and threw open the doors of the church, exclaiming, 'Maister, the cider do tuzzily!' Whereupon the rector hastily abandoned his congregation mid-sermon and rode home at the gallop to see what was going on with his cider. It may well have been a secondary fermentation. *Tuzzily* means foaming or mantling at the bung hole. In Somerset dialect 'Tuzzy-Wuzzy' is the burdock plant.

Thomas's son, Revd Thomas Merton Cornish, took over the living and ran the cider business from 1840 to 1856. By 1842 the reputation of Heathfield cider had spread far and wide and small barrels were sent to Queen Victoria, Viscount Melbourne, the Duke of Bedford, the Marquess of Worcester, the Rt Hon. Earl Nelson of Trafalgar House, Salisbury, and Lord Strathmore of Glamis. As well as to London, Cumberland, Dundee, Worcestershire, Yorkshire, Carlisle. Heathfield cider was even exported to Australia by 1852.

The next vicar, Revd Edward Bryan Combe Spurway, arrived in 1856 and continued cider making. His best cider was sold at 8 guineas a hogshead: cheaper at £1 7*s*. a hogshead. Spurway was vicar for forty years. A good innings. His son, Edward Popham Spurway, then took over both church and cider making. He used a hydrometer and was obviously a perfectionist. In 1901 he won a gold medal at the Bath & West Show, the only cider medal ever won by a clergyman. He was a fine cricketer, good batsman and wicketkeeper. He died in 1914. Sadly, three of his sons were killed in the First World War. The next vicar, Guy

Hockley, was 'a saint, but no cider maker'. No cider was made during the First World War and all the barrels and equipment were dispersed. Very sad.

In 1911, they lost their cider maker, Arthur Moore. He was offered an extra shilling a week and decided to go and work for George Vickery, a former cinema proprietor, at a new cider mill at Norton Fitzwarren. And that is how Taunton Cider started.

Another Somerset cider company, Coates of Nailsea, was founded by Redvers Coate, born in North Petherton in 1901. He had gone to school in Malvern, good cider country, and read chemistry at Bristol University. His 'intended' was Mary Catlow, whose father was yet another vicar, Revd William Edgar Catlow. About a dozen years later, over Sunday lunch, the vicar suggested to his prospective son-in-law as they were drinking a glass of cider, that he saw no conflict between God and cider making.

'You are an educated fellow, young Redvers but as far as I can see you have no job to go to. Why don't you start a company to make this fine drink that you and I are sharing?' Not bad advice from a vicar and headmaster. Redvers Coate and Mabel Catlow were married in Wellington in 1925, not far from Heathfield Rectory and Sheppy's Cider farm. 'Coates comes up from Somerset where the cider apples grow . . .' was a famous jingle in the 1960s.

Yet another cider company with church leanings began its life in the rectory at Credenhill, just west of Hereford. Credenhill is famous for at least three things: Thomas Traherne, a Hereford lad (1636–74), poet, clergyman, theologian and religious writer who was, after a spell in Oxford, rector at Credenhill for ten years (1657–67). His views on cider would have been interesting. His books are still widely read. At the core of Traherne's work is the concept of 'felicity', that highest state of bliss in which he describes the essence of God as a source of 'Delights of inestimable value'. Traherne would almost certainly have known John Beale, who was living in Hereford at the same time. Maybe his cider was of 'inestimable' value.

At the other end of the spectrum are the SAS - the Special Air Service who often work on special missions behind enemy lines. Their barracks are now at Credenhill. Here they learn, amongst many other things, not just unarmed combat and how to drink cider but the even undertake undercover reconnaissance in cider orchards. Hand picked late at night. 'Who dares to make cider, wins . . .'

And, thirdly, the rectory itself, which housed Revd Charles Bulmer and the Bulmer family. Revd Charles (1833–1918) was born in Hereford and was chaplain to the county asylum. Charles's family were Hereford wine merchants. He won prizes for bottled cider and perry. His wife Mary – Mary Grace Parnell Cockrem (1834–1925), daughter of Edward Cockrem, a Protestant Irish newspaper owner – was a Devon girl from Torquay. Her father was an interesting man: printer, bookseller, stationer, music seller and library owner. No doubt Mary had some of his Irish flair which their sons inherited.

Mary married Charles Henry Bulmer in 1862 and died aged 91 in 1925. She gave her second son Percy some good advice. He chose the business of cider making for two reasons. 'Firstly because my dear mother advised that the business chosen had better have something to do with eating and drinking because as she said "These things do not go out of fashion", and secondly because my father, the late Revd C. H. Bulmer, who was Rector of Credenhill for almost fifty years, was greatly interested in everything to do with the land and wrote a large part of a great work on apples and pears called the *Herefordshire Pomona*.'

In 1883 there were 23,000 acres of orchard in Herefordshire. By 1896 this had crept up to 26,000 acres – the same as Devon. Somerset had 24,000. In 1887, Percy at the age of 20, who had bad asthma, started making cider with fruit from Credenhill Glebe orchard. He used an old stone mill belonging to a neighbour, Mr Richard Whiting of Magna Castra, and a pony called Tommy. Percy's older brother Fred Bulmer came down from Cambridge to help. He was at King's College and later ran the business. Their father helped them to find premises and in the first year they made 4,000 gallons. It was hard work and the hours were very long. They could not afford a steam engine and hydraulic press. That came later.

By 1888 they were in Ryelands Street, Hereford, and the rest is history. In 1890 they produced Bulmers No. 7. Bottled and extra dry. They also made several trips to France: Normandy, Rouen and the Champagne region. With Thomas Traherne in the background, there was always added mystique, something Fred's son Bertram Bulmer often talked about. God and cider – a mystical business.

9

TWENTIETH CENTURY CIDER

At the beginning of the twentieth century cider was at a crossroads. Farmhouse cider makers still supplied cider to farm labourers as they had done for centuries. But then others working on a more industrial scale supplied the mass market in towns and cities. Factories, coal mines and docks were thirsty places. These two paths slowly diverged, until at the end of the twentieth century both ciders became two very different products with very little in common apart from the name 'cider' and the fact that they used 'apple juice', but in radically varying strengths.

Some even used less that 10 per cent apple juice. Why so? Different approaches, different philosophies. One was reliant on integrity and quality, the other on glucose syrup and apple concentrate. Very different beasts living in very different stables alongside each other. Or, as one well-known cider maker once said to me when we were judging cider, 'Hello, James, are we judging for quality or marketability?' I asked if they were not linked in any way? John Thatcher smiled. He was quite right; he had hit the cider nail on the head. It is all a matter of scale and approach.

Maybe it is worth looking more closely at cider philosophy. The way in which cider is made, marketed and sold often defines the cider philosophy for that county, that area or even that particular firm or farm. These days massive cider companies call the shots, yet they often rely on cider's rustic image and rural philosophy for marketing. Their ideas, which started out well at the beginning of the twentieth century, have been diluted down over the last fifty years. Of course they need apples but how much cider apple juice is in their cider? In large cider factories real cider wisdom is in very short supply. It must be kept in an invisible safe in the furthest warehouse, labelled 'Top Secret'.

Artisan cider philosophy was what interested me most of all. It produces cider that is exquisite and yet unpredictable. Cider, like fine red wine, is often better left for a year or two, an ethos in direct contrast to industrial cider made all the year round at high speed in a matter of days. Quality versus quantity. The old equation. A balancing act between reality and image.

After the Second World War what saved the day was a small band of independent medium-size, family-owned cider makers who stubbornly kept going. Most were in Somerset. Many of these families are still in business. Sadly, Devon lost 90 per cent of its orchards, and in Herefordshire and Gloucestershire many small and medium-size cider makers were bought up by Bulmers or just simply closed down. The basic laws of economics led to big cider firms running a monopoly type war, battling it out against each other with advertising and promotions, underpinned by the use of apple concentrate and newfangled methods of production. Cider was sold in 2 litre plastic bottles. BOGOFs. Buy one, get one free.

The dedicated hard core, 'long in the tooth' cider makers in Somerset were firms like Perry's, Sheppy's, Burrow Hill, Hecks, Roger Wilkins, Thatchers and Rich's. All still around, a rare breed. They kept the flame alive during the 1970s, 1980s and 1990s. They stuck to their guns making cider for four, five, even six generations. But cider making was often only one part of a working farm and that was crucial for their survival. Acres mattered. Individual diversity. Self-knowledge: that is another vital part of cider wisdom. Some like Thatchers have expanded out of all proportion but are still family-owned, just like Westons in Much Marcle. To survive these days cider makers have to be skilled in the subtle arts of negotiating contracts with supermarkets. Distribution and shelf space. A far cry from the world of straw and apple pomace.

But the increased quality of cider across the West Country at the beginning of the twentieth century did not just happen on its own. Far from it. There were one or two dedicated scientists and enlightened landowners working behind the scenes. Their cider philosophy was what mattered most. Science came to the rescue. Without their inventive work cider making would have disappeared and there would be no drinkable cider today.

Squire of Butleigh

Looking back with a long lens, Long Ashton's contribution to cider cannot be stressed too highly. LARS – Long Ashton Research Station – was originally called the National Fruit and Cider Institute and kicked off in 1903. The philosophy was very simple: improve cider making at all costs. Various notables at the Royal Bath & West Show like Robert Neville-Grenville, squire of Butleigh, and Charles Radcliffe Cooke MP from The Hellens, Much Marcle, in Herefordshire, thought that farm cider was pretty awful and that with a bit of basic science and logic it could be greatly improved. Charles Radcliffe Cooke left this damning comment: 'Ciders and perries are all alike and only to be distinguished from vinegar by a highly discriminating palate.' Pioneer work in apple identification had been done by the Woolhope Club in 1884 with the publication of *The Herefordshire Pomona* with its beautiful illustrations. But scientific research was now needed to help cider making.

The driving force was Robert Neville-Grenville who realized that good cider would vanish unless something radical was done. In the 1890s he used his own farm in Butleigh, near Glastonbury, to research cider making. A man well ahead of his time, Neville-Grenville concerned himself with drainage and agricultural improvements – he even invented a marvellous steam carriage. Neville-Grenville funded all the research work and employed one Frederick J. Lloyd for ten years to look into the science of apples. They set up laboratories in the stables, analysing cider apple juice and looking at fermentations.

Frederick J. Lloyd was a pioneer bacteriologist who had been employed by the Bath & West to look into cheese making. The Bath & West was more than just an agricultural show. It had been founded in 1777 and was a bit like the Royal Society for West Country farmers. It encouraged agricultural research, published learned papers and was always trying to improve breeds of cattle, sheep and arable crops, as well as swedes, turnips and cider. Competitions and exhibitions were common.

Lloyd studied at Bristol and lectured at King's College London. In the 1880s he had worked with Professor

Fig. 9.1 'Robert Neville-Grenville, Squire of Butleigh in Somerset, looking very dapper', 1906

Voelcker, an eminent agricultural chemist from Frankfurt and chemist to the Royal Agricultural Society of England. This osmosis of ideas and no-nonsense German approach was crucial. Chemistry was vital for understanding the hidden chemical and biological processes which were part and parcel of good cider making. Chemistry helped explain where things went wrong.

F. J. Lloyd also travelled with Robert Neville-Grenville to France in 1895 to visit the Pomological Society of France where they were both awarded medals. Lloyd wrote thirteen technical reports on cider making, which so impressed the Ministry of Agriculture that they looked favourably upon the idea of creating a National Fruit and Cider Institute. That is how Long Ashton Research Station was born.

F. J. Lloyd measured acidity, specific gravity and tannins in cider apple juice. He also made single-variety ciders, noting down the names of all cider apples that came within range, things we take for granted today and the lifeblood of good cider making. Thank you, F. J. Lloyd and Robert Neville-Grenville.

Long Ashton

Everybody was so enthused by the work at Butleigh that Lady Emily Smyth of Ashton Court leased them 13½ acres plus a few old buildings, including a cart shed and pigeon house. The lease was for 100 years and only ended in 2003. The location was well chosen. Long Ashton, near Bristol, was to become cider central for West of England cider makers.

Science was well on the way to explaining the universe and the atom, so why not the cider universe as well? Early researchers at Long Ashton were doing exactly that. Applying science to cider. F. J. Lloyd became the first director and in 1905 B. T. P. Barker, known to his friends as Bertie, became assistant director. In 1912 Long Ashton was integrated into Bristol University. Professor Barker worked there till 1943. He became one of the cider 'gods'. Long Ashton blossomed and quickly outgrew the original cider house and laboratory. Many of the new buildings were purpose-built red brick and extended over quite an area with laboratory blocks named Barker, Hirst and Wallace. Like a small village.

In a paper written in 1953, Professor Barker paid tribute to the pioneering work done at Butleigh by F. J. Lloyd and called it the real birthplace of Long Ashton. Hallowed ground indeed. A place of pilgrimage. Not just test tubes and hydrometers. A shrine to good cider without which many cider makers would have been completely lost. Cider chemistry was a wonderful world. The Greek natural philosophers would have been impressed. Eureka cider?

What then happened was public engagement. Farmers were encouraged to send in cider apples to be identified. The Research Station then made single-variety ciders, judged them, tested them and awarded prizes. Between 1903 and 1910 the juice of over 2,000 apples was analysed and from this data Professor Barker made four categories that stand the test of time. Sharp, sweet, bittersweet and bittersharp.

This was something farmers understood all too well. Long Ashton tasting days were famous. Farmers regarded it as a big day out. They came to see what their cider apples had made. Blends were made in the Long Ashton cider house in a uniform way. Some farmers went on courses to learn how to make better cider on their farms. Educational in the broadest sense. It helped to develop their palates. A vital ingredient. But they liked a party. Tasting days. Every year hundreds of people turned up, whole families in their best clothes and charabancs aiming for the cider marquee. By the 1930s the number attending was over 1,000.These tasting days became legendary. Large quantities of cheddar were ordered. One bill was for 3 cwt of cheddar cheese. How they ever got home one can only guess. Cider was definitely on the way up. Daniel Defoe would have approved. Sedgemoor cider?

Order was very slowly imposed on nature's chaos. Others might call it biodiversity. Scientists had been set a seemingly impossible riddle within an enigma which only DNA could solve conclusively. But the discovery of DNA was a long way off. Many cider apples had their own names which varied from village to village, farm to farm. Many looked similar. It was a hell of a job to disentangle them all, to see the orchard for the trees. Many wildings, crabs, seedlings, synonyms and varietal names. A thousand years of cider making had left its mark on the countryside.

Another person, apart from Andrew Lea, that I talked to about Long Ashton was the pomologist Liz Copas. She was very much up to speed with the history of LARS and was full of admiration for all the past work that they had done. 'Long Ashton? I fell in love with the place. All that fruit and cider. They did all sorts of things with blackcurrants. Vernon Charley did his work for Ribena between the wars. When I went there they were working on top fruit: Cox and Bramley. Viral control: finding out what the old trees were suffering from. And then finding methods of eliminating viruses and propagating new material, clean fruit. The sort of thing Thomas Andrew Knight was touching on. He did not know what viruses were but he had an intelligent attitude towards the problem.'

As Liz says, 'My boss was Ray Williams. Famous for his work on perry pears. A Forest of Dean man. He came from Longhope near Mitcheldean, not far from where they developed the Blaisdon red plum. He spent a lot of time in the 1960s and 1970s going across the Aust ferry on his bicycle and into the Forest of Dean looking at pears. He spoke the language.

'One good story was that Thomas Andrew Knight labelled as many perry pear trees as he could and identified them with small lead labels that he hung on the trees. At first Ray Williams couldn't find them until he was told that Dr Herbert Durham was on a horse. Ray was only on a bicycle. He looked up and there they all were, ten feet off the ground. Ray wrote the definitive perry pear book in memory of Prof. Barker. He also helped with the development of bush orchards. They were started in 1930s but never caught on. But when they started television advertising they suddenly needed more apples and pears quickly.

'My job was quite autonomous really. We did what we liked and had a free hand so long as we got the work done. My job was half and half. Half on pollination for Ministry work and then half on orchards for companies like Bulmers, Taunton Cider and Showerings. Main thing was to control biennial bearing, because all the bush orchards were all the same age and were giving massive crops one year and nothing much the next year. Not good for factory cider.'

Sadly, Long Ashton was targeted by Mrs Thatcher in 1980s. She was a grocer's daughter from Grantham and not a cider drinker. Her father, a tobacconist and Methodist lay preacher, ran the corner store. Cider would have been frowned upon. So Methodism in the end got the better of Long Ashton. But Long Ashton fought back. Some scientists clung on for twenty years until the lease ran out in 2003. A full century of cider research.

D-Day cider

While all the cider research was going on at Long Ashton there were two world wars. Large numbers of farm workers went to the front in the 1914–18 war and many never came back. That dented farm cider badly. One centenarian cider maker I knew was Jim Webber of Stoke Abbott, West Dorset. When I asked Jim when he was born he said, 'After the war.' I then asked, 'Which war?' 'Boer War.' He was born in 1902. I then asked Jim when he started work. 'Before the war.' The First World War. His father was 'sent to France with horses. Saw action and then came back.' 'Didn't say much about it.' So in 1914 Jim started work aged 12 ploughing with a team of his own. I wonder how many 12-year-olds could manage a team of horses today.

Jim was a hard worker and once cut nine acres of barley in seven hours, thirty minutes with a horse-drawn reaper binder. Quite a record. They needed cider on a hot day to 'oil the joints'. Lubrication was vital. 'Your throat got powerful dry.' Sometimes the farmer brought the cider out and sometimes they made their own. Manor Farm was where he had his old wooden cider press. That was where he made his D-Day cider. I saw the press many times.

In 1943 American troops turned up on their doorstep doing military training before the invasion of Europe. They caused a fair bit of mayhem in village halls with local girls and all that jazz. Troops based in Beaminster were from the 16th Infantry Regiment. They had seen heavy fighting in North Africa and now drank a fair bit of Jim's cider to prepare themselves for Normandy. Jim's cider was very popular. The local landlord wanted more, so in the autumn of 1943 Jim, with his brother Jack, bought in extra cider apples and barrels, then worked flat out in the evenings till Christmas. They made double the normal quantity: 15–20 hogsheads. Then in June 1944, when the cider was at its best, the American troops suddenly 'disappeared overnight'. A deathly silence fell upon the countryside. 'D-Day – no troops to drink the cider, see.'

So Jim was one of the very few people in Dorset to disapprove of the timing of D-Day. 'Should have had a word with Eisenhower when he visited Beaminster.' Sadly, most of the troops who drank his cider were killed on the very first day. They landed on Omaha Beach, taking very heavy casualties. Robert Capa, the war photographer, went ashore with them. Those soldiers are in his pictures.

Apple trains

To get another angle on cider in the 1960s I went to see a lorry driver and squeezebox player in north Dorset called John Cluett, he was born in Sturminster Newton in 1932 and was related to William Barnes the Dorset poet. As a boy he remembered Canadian soldiers from Newfoundland billeted in tents next to his farm. They knew what cider was and how to drink it. John was fascinated to hear them using Dorset dialect words, no doubt left over from cod fishing a century or two earlier. Plenty of Cluetts in Newfoundland.

In the 1940s cider was still a force to be reckoned with. Even on the front page of the *Western Gazette* there were adverts for farm workers: 'Good farm cottage and cider', even though the Truck Act prohibiting exchange of labour for cider had been passed in 1887. John knew of certain farms where high-octane 'apple wine' was made during the war. At 87, John still has a strong Dorset accent and still makes cider. 'The squeezebox do like a drop every now and then.'

We met for lunch. John was very cheerful. He wore a coat of many colours, a wide-brimmed hat and clutched a large hazel stick. He had worked on the Somerset and

Dorset railway line back in the early 1960s. 'We used to get apple trains come up from Hamworthy near Poole – bound for Bulmers in Hereford. Cider apples came over in a boat from Brittany and they were unloaded with a grab and dumped into twelve-ton steel coal wagons. Normally they took coal down to Poole gasworks. Hosed out, if you were lucky, coal dust no doubt added a bit of colour and depth to the cider. They loaded these wagons up with French cider apples and then they'd come through after the last UP train at about half past ten at night.

'I were signalman at Stalbridge, that was good, that was overtime. We knew they were coming up after the mail train. Around three hundred and fifty tons of cider apples on each train. Imagine that. One train a night for a week. Near enough two thousand tons of apples. A hell of a lot of cider apples, they always used to come up when the clocks went back in October. The smell was beautiful, just like cider and under a full moon – magic. The scent of French cider apples lingered in the night air for half an hour or more.

'Mind you we had our own cider works at Shillingstone, Lujen Robins. All the railway gangers would go in there. They always found something that needed doing near his cider shed. They used to fill up jars and take them on. They also had a little ganger's hut near the yard in Sturminster and that was like a cider den. You didn't dare go in there. You'd never come out again. You wouldn't walk out. You'd stagger. No bad cider just that some is better than others.'

Then John worked with a coal merchant and they would make clandestine late-night runs out to a cider farm near Stoke Trister and only just get back in time for milking.

In 1985 my brother and I discovered Lujen's cider den. 'Up for Sale.' It had been left just as it was when the railway closed. Untouched with hogsheads and pipes, presses, bicycles, cobwebs and wheelbarrows. A rare survival. For a few years we made cider in Durweston with a Victorian horse-drawn cider press and scratter we found in a barn when out sheep shearing. Every man must have a scratter to crush the apples. It was just like Worlidge's ingenio, only we fixed it up with a drive belt to a 1950s Fordson Major. Those were the days . . .

Factory cider

Farm cider staggered on, but only so long as there were farmers still alive to make cider and drink it. But much as the lights were going out all over Europe in 1914, so cider farms slowly went out of business one by one as industrial cider flexed its muscles and people left the land. In the late nineteenth and early twentieth centuries cider companies blossomed. They were ten a penny. Bulmers, Taunton Cider and Coates were only the tip of the iceberg. Gaymers in Norfolk started making cider in the eighteenth century as farmers, victuallers and publicans. William Gaymer kept the Crown Inn, Banham, and it was his son, also William (1842–1936), who radically changed the business. In 1870 they used hydraulic presses and in 1895 they moved into a factory site in Attleborough. Industrial cider making with a railway line close by. They expanded rapidly and employed 400 people. In 1940 its sheds, next to a railway line were assumed to be a prime military target, and were bombed with incendiaries by a lone Luftwaffe pilot. Gaymers were out of action for several years. It was the only cider factory to be hit during the war . . .

In Herefordshire, Symonds of Stoke Lacy started making cider in 1727 and continued with 'Scrumpy Jack' until 1984, when they were brought out by a brewery. Then they were sold to Bulmers. Stoke Lacy works closed in 2000.

Westons of Much Marcle have survived as a family-owned cider firm. They were started in 1880 by Henry Weston who was born in 1850 in Cheltenham. His family, originally from the Forest of Dean, were linen drapers, a clean-cut business dealing with the public and a quality product. Aged 11, Henry was found to be living with his aunt and uncle at Upton Bishop. By 1878 he was a tenant farmer at The Bounds, Much Marcle, where the cider works is to this day. Right from the start he wanted a quality product and stuck to his guns. Also living in Much Marcle was the MP for cider, Charles Radcliffe Cooke. Westons have many old large oak vats. Pip, Squeak and Wilfred are 24 feet high and hold about 42,000 gallons each. Henry Weston did not believe in advertising and thought that cider should speak for itself. Just down the road is Dymock, where several famous poets lived and drank cider in the summer of 1914, Edward Thomas, Robert Frost and Rupert Brooke among others. They also spoke for themselves. 'Adlestrop'.

'The Road Not Taken'. 'Is there cider still for tea?' Today Westons is a multi-million-pound cider firm with over 200 staff run by Helen Thomas, the fourth generation of the Weston family.

Another large cider firm, Whiteways in Devon, was started in 1891 at Whimple, 8 miles east of Exeter. They moved from Harbertonford, near Totnes. Like Gaymers they were near a railway line. Railways were important. Within a few years they were sending cider to the Far East. In those days the desire to keep the quality of cider worked to their advantage. Other well-known Devon firms were Henleys of Abbotskerswell, Symons of Totnes, Hills of Staverton and Inch's of Winkleigh, who were bought by Bulmers in 1995. All had their following. Cider was big business. Exports were on the horizon.

In the first half of the twentieth century the philosophy of industrial cider was close to that of farmhouse cider. It was just on a much larger scale and had better distribution and advertising. Orchards were booming. Competition, sales, dominance, market forces and bottom lines were key words in the boardroom. Bulmers had a fair slice of the Midlands and Wales and sent cider to Ireland. Whiteways and Westons, London and the south; Taunton Cider had Somerset and Bristol; Gaymers had East Anglia and the LNER line towards Scotland. One classic way to secure your customers is to buy out the local opposition. So that was when many good medium-size cider makers were bought out.

Cider was fine between the wars. Then, after 1945, sales slumped. Cider was no longer seen as a quality drink alongside wine at the top end. It had slipped down the scale and was now competing with beer in public houses, even though it was cheaper and stronger. The laws of big business began to overtake the natural laws of farmhouse cider making. The cider market was fragmented. You make farm cider once a year in the autumn, but beer is made all year around. You can get malting barley from anywhere. Cider philosophy was changing fast. Industrial cider became like any other business. It competed with beer, not wine.

In the 1950s the shape and feel of the cider counties was also changing. The 1953 *Bottlers' Year Book* lists 63 commercial cider makers. Somerset had the largest number with 19, followed by Devon 13, Gloucestershire 8, Herefordshire and Kent neck and neck with 7, Norfolk 3, Hampshire, Worcestershire, Hertfordshire, Surrey, Suffolk and Northern Ireland all with 1 each. Bulmers was already buying up rival companies. Takeovers were common.

Somerset had Showerings who started Babycham, a sparkling perry which proved immensely successful. 'Mine's a Babycham.' They bought out Coates and Gaymers and were themselves bought out by Allied Breweries. Beer triumphing over cider . . .

Medium-size cider farms in Herefordshire and Gloucestershire were hoovered up. Symonds in Stoke Lacy sold out to a brewery and then Bulmers got hold of it. Inch's in Devon went to Bulmers then reinvented itself. Merrydown left the main stage in 2004 and their cider-making is now contracted out. Taunton was run by brewers and various consortiums for years, then bought out by

Matthew Clark, then Gaymers and Magners. Very complicated. Bulmers was bought in 2003 by Scottish & Newcastle. Then in 2008 they were bought by Heineken, the Dutch lager company. So cider, which in the 1650s flexed its muscles alongside French wines when the Navigation Acts prohibited trade with Holland, is now in the same stable as Dutch lager. Ironic.

Brewers buy up cider makers. Never the other way round. Even Aspalls Cider in Suffolk, founded by the Chevaliers in 1728, who had brought their granite Chausey cider presses from Jersey, were bought up in 2018 by Molson Coors, the Chicago based brewers of Miller Lite.

But I wanted to find out what had precipitated these shenanigans in the industrial cider-making world back in the 1960s. This was the time when several rather dubious 'innovations' were enacted behind closed doors. To find out more I turned to a well-respected cider expert with an enquiring mind. Andrew Lea had just come back from researching cider apples in Kazakhstan. He had also worked at Long Ashton Cider Research Station.

Cider chemist

Andrew Lea is a biochemist with a deep interest in cider tannins. He lives south of Oxford, near Dorchester-on-Thames, an ideal spot for orchards. Andrew always wore his trademark light blue sweater. I had known him for nearly twenty years. We had judged cider together several times at the Bath & West and in Hereford. A very pleasurable experience. It is only when tasting cider and discussing its virtues for hours on end that you develop a finely tuned palate, plumbing depths and complexity. Intuition honed with a lifetime's experience. I knew I had come to the right man . . .

How had Andrew entered the colourful world of cider? He is a quiet man with a finely balanced sense of humour. Both his parents had been in the navy during the war, his father on convoys and his mother in the WRENs. They met in Naples, close to Vesuvius where Pliny came unstuck during the famous eruption of AD 79. Not a bad spot. In Pompeii archaeologists restored the House of the Orchard with its superb murals. Another large orchard of over 100 trees has also been discovered, and in the middle of it an altar for offerings to Pomona. The gods were smiling.

Andrew grew up in north London and studied chemistry at Bristol University. His first job was with Brooke Bond Tea, near Reading, looking at tannins in instant coffee. Instant coffee. Instant tea. Instant cider. A nightmare waiting to happen. Add water and stand back. Then fate, like Vesuvius, played a hand. His wife Jo got a job as a GP in Bristol, so they moved there.

By chance Andrew met Colin Timberlake at a tannin conference. They chatted and Andrew realized that tannins in cider were very interesting. Colin worked at Long Ashton and soon afterwards Andrew took up a PhD studentship there. This was in 1972. Five years later his thesis appeared: 'The importance of phenolics and tannins in cider apples and cidermaking', a vital part of understanding the biochemical processes we often took for granted. 'It is tannins of course that give the characteristic flavour and colour that makes good cider and good wine.'

'What had not occurred to me was that in those days Bulmers and other cider companies were afraid that cider production was so outstripping their orchards that they couldn't get enough bittersweet cider apples to fulfil the demand they had created. And there would be a lack of tannin in the cider.

'So they were worried that they wouldn't have enough high-tannin apples, which was very ironic because within a few years the marketing people had completely pulled the rug from under that. They persuaded the punters that they would be much happier with less tannin.' In other words a lighter and less challenging cider. No muscle. So they used arrows and archers as part of their advertisements, trying to beef up the macho image of their cider while lowering the apple juice content and tannins at the same time. Very cunning.

'So in the end the problem sorted itself out in an unexpected way. Tannin was my entry into Long Ashton and the world of cider. I have been in tannin almost all my life.' Talking to Andrew about cider is always a pleasure. Such breadth and depth of knowledge.

Cider tannins

Andrew Lea was at Long Ashton for thirteen years. The research buildings were at the far end of the long, rambling village just where Wild Country Lane meets the old Weston road. I wanted to learn far more about tannins, these powerful chemical compounds that affect the taste of food and drink. But what are they exactly? All I knew was that tannins had something to do with tea leaves, oak bark, tanning leather and the treatment of burns prior to the Second World War.

Tannins also helped make a good, even excellent, red wine. They give dryness, bitterness and astringency. The secret is in how the tannins mature. Without tannins top-notch Bordeaux would be nothing. And the same could be said for Somerset and Herefordshire cider. High-tannin ciders not only taste more complex than low-tannin cider but keep longer and develop far greater subtleties in the bottle. Long life in both senses.

Technically tannins are naturally occurring organic compounds called polyphenols, found in plants, seeds, bark, wood, leaves and fruit skins. Complex macromolecules made of oxygen and hydrogen molecules. You lose your tannins at your peril. You have to keep them, cherish them and understand them. Tannins also work as a natural antioxidant to protect the cider and even the cider drinker into old age. If you look after the tannins in cider they will look after you. Low-tannin cider is a bit like drinking Pinot Grigio rather than a full-bodied Merlot or Cabernet Sauvignon. But then again, many ciders are blends of maybe a dozen different varieties of cider apple, and so chemically they are often far more complex than wine. A fascinating world still waiting to be fully explored.

Tannins help age wine and cider, though few cider

makers deliberately age their cider. It's usually done by accident. I have had bottled cider, ten, fifteen, twenty years old and, in one case, forty-five years old: Horrell's Devon Cider. Bottles left in an orchard box at the back of a barn by mistake or under the stairs or in the cellar. It just depends how good the cork is. And very good Horrell's cider was, too. Deep, dark, golden brown and sweet, like a dessert wine. Excellent on a summer's evening. Just as I remember it as a boy.

Sadly, large cider companies have often turned their heads the other way for a quick buck and to stay in business. They pit diluted, low-tannin ciders against lager to the extent that cider has become little more than an alcopop with no more than one-third apple juice. The rest is glucose syrup, i.e. sugar, fruit concentrate, water and colouring. What a comedown for Lord Scudamore's Redstreak. He would be turning in his grave at Holme Lacy.

Andrew Lea tells me about tannins: 'Most fruits have tannin and there is a difference between tannins in pears and apples, though not as radical as tannins in oak bark and chestnut bark. In perry pears tannins are more polymeric and so they have a very different feel to them than the tannins in apples which tend to be of lower molecular weights, and that impacts on the way that tannins behave during processing. In some perrys you get a massive amount of tannins dropping out and they are on the edge of solubility in the fruit.

'Tannins affect the taste and flavour. Two primary aspects. One gives bitterness and astringency. What is called "mouthfeel", that gives you the body in the beverage and this is why high-tannin cider apples make far better cider than dessert apples. When all is said and done.' Bravo.

'The secondary effect of tannins is during fermentation in cider. The breakdown products are mostly bacterial and they can become volatile. That's where you get the bitter-sweet aroma. The bittersweet cider aroma isn't inherent in the fruit itself but forms during fermentation, when bacteria acts upon the tannins and pulls them apart. This liberates smaller more volatile molecules which are the source of the bittersweet aroma and emblematic of good cider. Fruity, oaky, apply notes.' This was absolutely crucial and gave me real insights into the importance of tannins in cider.

Andrew's work at Long Ashton also went in other directions to do with colour in food and drink because the whole food industry was looking at replacing artificial colours with natural colours. So they looked at extracting colours from natural sources like willow, whose bark can be very brightly coloured. 'None of the ideas actually took off at that time but many of those initial problems have now been overcome. We were well ahead of the game.'

The research work was incredibly wide. 'Prof. John Hudson was director. Fred Beech had just taken over as head of the cider section. Alfred Pollard had retired but he came in quite a bit. The cider section was renamed "Food and Beverage Division". We had good links with the wine trade in Bristol – Harveys and Averys. We did a lot of work on wine and port wine for Cockburns. Vernon Charley, who invented Ribena, had gone to Beechams after the war. He was research director. But of those scientists no more than half a dozen were working on cider. Most of the research had gone off to plant pathology, plant physiology and crop research. Horticulture. Even in pomology it was just Ray Williams, Robin Child and later on Liz Copas. Quite separate from the cider section. We hardly ever communicated except over lunch.'

So I came to view Long Ashton as the Bletchley Park of the cider apple world. 'It retained that government-funded research operations approach, plus a bit of secrecy. Very hierarchical. When I first went there everybody was called by their surnames. Prof. this, Dr that. Mr and Miss so and so. By the time I left we were all on first-name terms. Very civil service in lots of ways. A visiting group would come down from the Agricultural Research Council. They thought we were very ramshackle as we had no paintings on the walls and no carpets. They said they couldn't possibly judge how important we were. In other words they hadn't a clue about the value of the work we did.

'We were like one big family. They were happy days.' Tannin was vital for good cider and Long Ashton was vital for the cider world.

Dilution

But while Long Ashton was beavering away in the 1970s, large cider companies were tinkering with cider behind closed doors, gradually diluting cider down till it only had about one-third apple juice. No one batted an eyelid, apart from a few switched-on farmhouse cider makers. But why did the large cider companies veer away from traditional full-juice cider when it was so good, so deeply flavoured and rich in beneficial tannins? Why indeed . . .

Accountants and marketing men have, I suspect, much to answer for, never mind the breweries and their 'new' fiendish techniques involving glucose corn syrup made from maize, which was far cheaper than traditional malt, the starch being converted into sugar by acids and enzymes. This is a process invented in Russia in 1811 by German scientist Gottlieb Kirchhoff. By the 1970s the addition of glucose syrup to cider became common practice in cider factories.

Would the same thing happen in France with wine? I think not. When the widespread use of extra sugar occurred in the French wine industry in 1907 and cheap wine was imported from Algeria, there were widespread demonstrations and riots in Languedoc. They went on for months. At some gatherings in Carcassone, Beziers and Nimes over 250,000 people turned up. Twenty-two regiments of infantry and 12 regiments of cavalry were sent there. 25,000 infantry and 8,000 horsemen. Clashes were frequent. All because of extra sugar and cheap imported wine which undercut their business. People were starving.

At least six demonstrators were killed. One infantry regiment mutinied. 'Grapes for wine, sugar for sweets!' was one of the slogans. The riots nearly brought down the French government which quickly passed laws to prevent excessive use of sugar and wine fraud. They had some standards.

Laws in England preventing food and drink adulteration go back to medieval times, notably the Assize of Bread and Ale in 1266 which has been in force for 600 years. Bakers caught adulterating bread were punished very severely indeed. Adding chalk was not permitted for hundreds of years but nowadays by law flour has to have chalk in it. Calcium carbonate for bones. Times have changed. What was once regarded as adulteration is now viewed as smart business practice. Cheap bread and cheap beer no doubt keep everybody happy till they have almost forgotten what the original beer or cider or bread tastes like. Then the companies make clever little films about it. These advertisements try to reconstruct your memory, to convince you that you are drinking or eating the real thing. Like the Hovis commercial shot in Gold Hill, Shaftesbury, Dorset.

Changes in industrial brewing in the 1960s had a knock-on effect on cider, particularly as cider makers were slowly bought out by breweries. Andrew Lea, as always, had the answer from behind the scenes. A small, quiet voice in the midst of these vast cider shenanigans. To hear it from Andrew was an education:

'The veering away from bittersweet cider apples at that time was really a marketing thing. The beverage market had changed a lot by 1970. Harp lager, [was] about the first of the cold, fizzy, tasteless lagers (around 4 per cent). People were clamouring for drinks like that. The problem with cider was that young people didn't like these full-bodied, old traditional ciders. It wasn't cool. The marketing people wanted a cider equivalent to Harp lager that could be sold in bars up and down the country.'

Other factors were also in play. Between 1947 and 1960 more than two million national servicemen served in the army. Many were sent to Germany. A very boring way to employ young fit men. So the only consolation was to drown their sorrows in the Bier Keller with lager. Fighting the Cold War on the lager front. National service ended in 1960, the same year that Guinness pioneered Harp lager in bottles. In 1964 it came on stream as draft lager and was a great success, much to the consternation of the cider fraternity.

So the big cider makers deliberately made cider with progressively less tannins and less apple juice, intentional dilution which went undetected for many years, maybe reducing at 1–2 per cent a year so that nobody would notice. The big cider makers were notoriously cagey about dilution and concentrate. The only thing I could detect was that 'they had learned a lot from the brewers . . .' As the graph of the cider juice content fell away in a steady decline so the graph of their profits went the other way, upwards. They were able to make 3 gallons of cider out of one gallon of cider apple juice. Brilliant . . .

In 2004 a report was published by the Food Standards Agency which looked at 'Juice Content in Commercial Ciders'. Twenty-nine ciders and two perries were tested. The results were startling, to say the least. The variation was between 7 per cent juice (white cider) all the way up to 100 per cent, with most of the large commercial ciders being around 30–50 per cent. Five were below 20 per cent. Six were between 20 and 30 per cent. Six more were between 30 and 50 per cent. Seven were between

50 and 70 per cent and the rest were above that. A very wide variation indeed. Where were regulations? Where indeed.

Interestingly, a 1980s television advert for Strongbow set in a pub uses the final clinch line as *Strong, Straight and True*. When I asked one PR lady in Hereford about all this back in 2008, she simply said, 'James . . . You know I can't possibly tell you the juice content of our ciders. It is a closely guarded "state" secret. Our cider is made to a well-known, tried and tested ancient formula.' I got what I deserved. A cheeky question sometimes deserves a cheeky answer.

Sweeteners

Dilution was not the only thing going on behind the scenes. In the 1970s, as the UK was entering Europe, the NACM, the National Association of Cider Makers, abided by their own rule book – known as Customs 162. This is the official HM Customs & Excise document that is supposed to monitor the industry. What it says is alarming, if you are an artisan cider maker. Any amount of sugar can be added, as well as water and many other things. Apple concentrate, both home-produced and foreign, was welcomed with open arms by industrial cider makers. Trying to get the juice content of ciders put onto the label is an uphill struggle. Curiously enough, in the food industry there are far stricter regulations about disclosure. Even the famous Mars Bar has its main ingredients listed on the wrapper, including glucose syrup. If a Mars Bar made in Slough can list all their ingredients, why not cider makers?

My own belief is that the public has a right to know the juice content of ciders they are drinking and the levels of glucose syrup and added water within the cider. Gone are the turnip days of the late nineteenth century when beetroot added a bit of colour. But at least beetroot was natural. Might even have improved the flavour if dessert apples were used. Sometimes called red diesel . . .

Andrew again. 'When we entered the Common Market there was no definition of cider in the UK. So the industry put together a definition of cider in their code of practice and that became adopted as Customs 162 by the NACM. Both MAFF (the Ministry of Agriculture, Fisheries and Food) and HMRC [as Customs & Excise was later called] regarded the NACM as the authoritative trade body for the industry, they took the whole of cider practice as written by them as gospel.'

The real problem is that by slewing the market with inferior cider puffed up with macho advertising, many small and medium-size cider makers were either bought out or driven to the wall. It has not been a level playing field for over half a century. Today we are still living with the consequences of those rather dubious 'innovations'.

Not that long ago the NACM came out with strange statement that the juice content of their ciders would not fall below 35 per cent. It was closing the stable door after the horse has bolted. As Andrew Lea said again: 'as the graphs of tannin content and juice content went down over the years in one direction so the graphs of sales and profits for the big companies went the other way . . .' *Fait accompli*.

Concentrate and yeast

Plumbing the depths of the large cider makers' psychology was a bit like diving into a parallel universe where black was white. Like discovering another orbit, another planet, another set of rules. Cider ethics at that time was an unruly beast lurking in the shadows. Answers were far from clear. Artisan cider makers often wondered if there was ever a cider apple in sight. Some firms turned to white cider to save their fortunes but that was ultimately a lost cause and only led to bad publicity and death in the gutter.

Andrew Lea saw all this happening from the sidelines and my suspicion is that he was appalled by it, as all true cider makers would be. As Andrew says, 'By the time I was at Long Ashton the big companies were quite protective of their own research. They saw the future as just a few big companies. So we had very little connection with small cider makers. Which was strange as the original remit for Long Ashton was to help farm cider makers make better cider. And in most cases they succeeded. For a while the big three companies, Bulmers, Showerings and Taunton. had well-appointed research labs of their own.'

These days about 90 per cent of the cider sold in the UK is made with apple concentrate and has an apple juice content of around 35 per cent. So it is a vital part of cider history and economics. It is not a level playing field at all.

Orchards are rare territory for some cider makers. I have heard that there are laptop cider makers who have never pressed an apple in their lives. The virtual cider world - where you buy concentrate and fruit flavours over the Internet - is a very strange place. Cutting edge or cutting corners? Add water and stand back . . .

But what is apple concentrate? It is apple juice that has been reduced about 10:1 by evaporation, i.e. heating it up to boiling point with steam so that most of the water vapour comes off, with reduced pressure, to leave a dark apple syrup, which arrives in large tubs or a super-tanker depending on how much you want. There is always a kind of appley caramel smell to it, not unpleasant, but you wonder where all the really interesting subtle apple aromas go. Likewise pasteurisation helps preserve the cider but also dumbs the flavours down.

One of the real problems was that for many years the big companies would not acknowledge in public that they used concentrate or that they used glucose syrup and not apple juice as the main fermentable sugar. It was a major shift . . .

Andrew Lea again: 'The problem was that after the war the brewers realized that they could make beer with a lot of glucose syrup and not so much malt. I'm sure accountants loved it. "If we can do it with beer and no one is batting an eyelid, why can't we do it with cider?" The punters didn't notice. It fitted the *Zeitgeist* of having a lighter type of beverage. The Government didn't mind. They probably didn't even know. They were getting alcohol tax on water.'

But there was also a dark undercurrent of secrecy. The large companies were quite embarrassed to the extent that they were not really allowed to talk to the scientists at Long Ashton about it. Andrew had a strange experience on one of his visits to Hereford. 'I once mentioned these things to the technical director of Bulmers and he said, "You must never ever mention these things to me again. Never talk about these things. These things are closed. They are secret. You must not tell anybody else what actually goes into cider." It was a bit frightening in some ways, but it was the way the industry was changing at that time.'

Then there were yeasts, the vital part of fermentation. That was fascinating. Wild yeasts that are deeply engrained on the cider apples, on the equipment, on the cloths and hanging in the air at cider farms. Wild yeasts so beloved by small artisan cider makers, unpredictable and difficult to control on a large scale. So the big companies like Bulmers used cultured yeasts for many years. Dr Durham, their chief scientific adviser, had isolated a lot of yeast cultures in the interwar period which were very useful.

'Their main yeast was Number 18. In those days you couldn't just go out and buy dried yeast as you can today. They all had to be kept in a microbiology lab and sub-cultured. Taunton had a yeast plant. You had to grow it on and maintain it yourself. One of the things that Fred Beech and his group at Long Ashton were trying to do was to find a yeast which was low-sulphite producing, that was why we went to the Australian wine yeast AWRI 350. This was actually a French yeast that had been put into the

Australian wine yeast collection in Adelaide. A low-nutrient yeast. It had a lot to recommend it. AWRI 350 was the main workhorse for fermentation. It was all about control and consistency.'

These 'tricks of the trade' gave the big industrial cider makers a distinct edge economically. They gained consistency and carte blanche to expand. Over the years with brewers' distribution networks they effectively wiped out lesser companies who were often making a higher quality cider. Even Bertram Bulmer once declared, in jest, that you would not 'catch him drinking any of his own products', unless it was Bulmers No. 7 which was an old-style dry cider, deeply respected by the cider fraternity. They still talk about No. 7 today in Hereford with great affection. Industrial cider is a very strange world. They still need cider apples. But not as many as before and many bittersweet orchards are being cut down as we speak. Most industrial cider has now become very bland and uniform which is tragic.

Death of LARS

Sadly, all good things come to an end, but no one could have predicted the swift end of Long Ashton. It was like a bolt out of the blue. Mrs Thatcher was involved. Andrew Lea was as shocked as anyone. 'All this research was going along quite nicely and then in 1981 the Agricultural Research Council without any consultation stopped all research work. Fruit and veg, top fruit, apples, pears, blackcurrants, etc. They decided to switch from horticulture to agriculture, from fruit to arable crops. They certainly didn't want to subsidise cider research any more despite the fact they got significant tax revenues from it.'

'The Thatcherite principle was that 'near market' research should not be funded by the Government but by the industry itself. They saw horticultural support as excessive compared to the economic value of the sector but bizarrely they saw arable crop research as underfunded. Money should now be switched into arable crops where the petro-chemical fertilizer industry had a massive role to play'. Chemical farming v. chemical cider. 'How much the decision to close Long Ashton was politically inspired and how much was economic no one knows. You can tax cider but not potatoes.'

Doom and gloom. 'We had lost our *raison d'être*. We tried to increase the amount of contract work for MAFF but the writing was on the wall. Eventually the Food and Beverage Division closed in 1985. Most staff took early retirement. Cider pomology was kept going by Ray Williams with Liz Copas because the NACM was prepared to fund it.'

Andrew left Long Ashton in 1985 and was offered a job at Cadbury Schweppes (as it then was) in Reading. They had just bought an apple juice company in the US. 'I worked on apples and then tannins in chocolate and raw cocoa.' Lucky man. Tannin is an important ingredient in dark chocolate. Andrew was made redundant, but they formed a company for analytical work in the food industry.

'I picked up my old contacts in the cider industry. Some research work but mostly troubleshooting. I became more of a flavour chemist and developed a real taste for my work.'

In 2003 Long Ashton was finally closed and the site sold off for housing, an ignominious end to a hundred years of dedicated research into cider and orchards. And now, with the renaissance of artisan cider, the knowledge accrued at LARS is needed more than ever. A real shame Mrs Thatcher wasn't a cider enthusiast. She had after all been a food research scientist. Maybe she preferred emulsified ice cream. Cider-tasting days at Long Ashton are still talked about with great affection. It was a wonder any of the cider makers every got home in one piece.

The cider guru

Andrew Lea's feeling is that the best cider is made with minimal intervention: 'But the intervention you make should always be to make the cider more palatable. Not to extend it.' The large-scale cider makers have several different philosophies of their own floating around, philosophies often dictated by size and method of manufacture. The word 'manufacture' should tell you all you need to know.

One thing that has certainly changed over the last few years is that the large companies no longer want so many traditional high-tannin bittersweet cider apples. They want a fairly boring base cider often made with dessert apple concentrate either from here or abroad that can take a fruit aroma and essence. Fruit ciders are all the rage but are they really cider?

Maybe the artisan cider world should flex its muscles. People should at least know what they are drinking and how it is made. Anything less is fraud. On the plus side, over the last twenty years there has been a very lively renaissance of ethical, small-scale, first-generation cider makers often in their late twenties or early thirties. Whether they become viable businesses in the long term has yet to be seen, but I wish them well in every sense. They deserve to succeed. In the past West Country cider has often only survived by being part of a larger, mixed farming system. Maybe there is a middle path. Diversify. Become a chef. Drive a tractor. HGV cider.

Andrew's philosophy is that cider should be viewed on the same table as wine. He thinks it is unfortunate that in Britain and America cider has been seen as a variation of beer. 'Cider works better if it is viewed as a wine. High-status cider as it was in the seventeenth and eighteenth centuries.' History to the fore.

Andrew Lea's story doesn't end there. In fact it now enters its most interesting phase. When Andrew left Long Ashton he moved to a small Oxfordshire village and planted an orchard next to his house. He carried cider research out in his own backyard. 'Just a little hobby to keep a foot in both camps.' The day job was with big companies and at the weekends he had his own cider and orchard to attend to. His enquiring mind was never short of ideas.

Andrew started from scratch and became a small-scale

hobby cider maker. 'There were no books at the time that I thought were any good and there was a lot of misinformation.'

In the early 1990s information on cider making was very sparse. Andrew wrote a few articles for a self-sufficiency magazine, and when the Internet came along he put those articles up on a website. 'It began to mushroom.' Andrew was in demand, the unofficial small-scale 'cider guru'.

Andrew then took a radical step. Using uniform cultured yeasts was no longer interesting, so he went back to wild yeasts and 'thought how best to manage them, checking phs etc., not letting them go hell for leather. That was a novelty for me. You lose a certain amount of control but you gain so much and sometimes you get stunning cider. I found it was working for other people as well. Many had tried cultured yeasts and were disappointed, and then they started using wild yeast. Apples come with their own wild yeasts on their skin. Why meddle with what nature has given you?'

Andrew's enthusiasm and deep knowledge of cider biochemistry has helped many small cider makers all around the world and kick-started several businesses. His small handbook is a bible for those experimenting with cider making. In some areas Andrew Lea has cult status and has in his own way continued to help and support small cider makers, whoever they are, whatever their size. Continuing the ethos of pre-war Long Ashton . . . Squire Neville-Grenville and Bertie Barker would raise a glass to him.

Common Ground

The sad truth was that by 1980 most cider orchards and cider makers were keeling over. Their children weren't interested. They grubbed 'father's orchards' out for grants. Farmhouse cider making nose-dived dramatically. Developers did as much damage as lager, and the end seemed terminal. Even Heathrow was once a series of large orchards and market gardens. But as orchards were sacrificed to make way for new houses, something extraordinary happened in the national consciousness that made people more aware of apple trees and cider.

The first stirrings were in about 2000. The subsequent upsurge in artisan cider making was a grass-roots reaction against the industrial cider makers. In a sense it was a rebellion, a rebellion of the senses. People wanted high-quality cider, they wanted to know where the cider apples came from and what the apples were. As the desire for artisan cider gathered momentum the groundswell of public interest seemed unstoppable. Real cider re-emerged.

Apple Days boomed. All over the country people reconnected with the land and there was a bonanza of orchard planting. Community Orchards sprung up. Thanks to which, countless varieties of apple have been saved from extinction. Cider clubs are all the rage and wassail is back on the menu. But how did this happen?

You have to wind the clock back to the late 1980s. To find out more I went to see Common Ground, a dynamic, ideas based charity that was instrumental in kick-starting Apple Day and getting it into the calendar. So on a cold but bright February morning I drove to Shaftesbury. The A30 winds through honey-coloured hamstone villages. Hedges low – landscape laid bare – minimal and glistening.

My rendezvous was with Sue Clifford and Angela King – 'Old Common Ground' as I call them. Back in October 1990 they promoted the first large scale Apple Day, a massive event right in the heart of London that went on all day. They thought big right from the start. Not just the odd village; they wanted to encourage the whole country even inner cities to embrace the idea of Apple Day and run with it. Very wisely they chose London's old fruit and vegetable market. Many thousands descended on Covent Garden that Sunday morning to get a slice of the apple. Apples were topping the bill, strutting their stuff on stage. People came from all over the country. British Rail even lodged a complaint that they had not been forewarned. It was like a football match.

Almost all other apple days around the country have flowed from that one event, and abroad as far afield as Tasmania and America. Subsequently, Common Ground inspired many other projects to help raise awareness of the commonplace, the everyday things, not just the rare and endangered. They worked with artists, poets, writers, naturalists, filmmakers, composers, musicians and sculptors. One year I was employed as Apple Day poet laureate. Brilliant.

Straight to business. A quick cup of strong coffee and a small croissant. I wanted to find out how the idea of Apple Day was conceived and the philosophy behind it. Where did the idea come from? Why did it work so well? Angela

King sat quietly at the far end of the oak table. Angela was the campaigner with the background in activism. Sue Clifford was the writer and animated communicator. Founded in 1982/83, Common Ground was small but very effective. They evidently had clout. They inspired people. Also on board was writer and filmmaker Roger Deakin, he who later swam in wild rivers and moats and wrote *Waterlog*. Very quickly Common Ground became a powerhouse of new ideas. They hammered on doors and slowly changed the way in which we perceive the rural world. They concentrated on habitat conservation and everyday ecology.

From their office in Covent Garden they produced many ground-breaking publications which showed people how to appreciate what was on their doorsteps and fields, and spurred people into action where needed. Not an armchair in sight. In 1984 Common Ground produced *Second Nature*, edited with Richard Mabey which included essays by John Fowles, Fay Weldon, Edward Blishen, Peter Levi and illustrations from Elizabeth Frink, David Hockney and Henry Moore.

Common Ground was afloat with ideas. Then they stumbled upon an amazing fact. 'We learned that there were hundreds of different apples and that astounded us.' Not just hundreds but about two thousand. At long last orchards and apples hove into view like a tall ship on the horizon. Common Ground had found its apple *métier*, its *pièce de résistance*, its core value and *raison d'être*.

The real catalyst was the Great Storm of 1987. It caught

everyone by surprise. Tragically, many trees were lost and many orchards were uprooted, particularly in Kent. Centuries were felled in a single night. The hurricane swept all before it. Half a million trees keeled over.

But Common Ground did something rather curious and low key. They distributed 56,000 postcards, featuring illustrations by David Nash, and encouraged people to let nature take its course, not to clean up after the storm too hastily. They saw the Great Storm not just as wholesale destruction but an opportunity for nature to reassert itself, to regrow according to its own rules. Their key emphasis then switched to orchards.

SOS

Common Ground then adopted a new ethos. In September 1988 their campaign 'Save Our Orchards' was started. A Mayday call. An SOS. At around this time another campaign also called 'SOS' emerged in Bristol Docks led by a flamboyant cider maker called John Dix, also known as 'Dixie'. He went round to shows selling cider from a thatched cider lorry. His SOS meant 'Save Our Scrumpy'. He had broadsheets printed which are now collectors' items, extolling the virtues of full-juice wild yeast fermentation from cider apples. Both campaigns ran independently but in parallel. Both worked very well.

Save Our Orchards had a leaflet, just like old-time pamphleteering. Well-printed, well laid out. People could take it home and inwardly digest. Ideas seeped into the

Fig. 9.2 'Nigel Stewart of Bridge Farm Cider, explaining the finer points of cider making', Covent Garden, Apple Day, 1990

nation's consciousness as the pamphlets were read over the kitchen table.

Common Ground then commissioned James Ravilious to travel round Devon, Dorset and Somerset photographing orchards and cider making. They wanted to inspire a new way of looking at orchards and conserving them. They worked with the Countryside Commission. James was the son of war artist Eric Ravilious. He inherited his father's artistic eye and knack for capturing the essence of rural life. He threw himself into his work. He was brilliant.

Ravilious's black and white photographs were exhibited in Exeter in October 1989 and toured for several years in village halls, libraries, museums and art centres. I once spent the best part of a day with James chatting with him at his home in Chulmleigh in Devon. I had great respect for his work and his appreciation of the odd, peculiar and commonplace. We talked about orchards and cob barns, straw, sheep, cattle and cider. When you work on a farm you notice the same sorts of things. James's photographs have a timeless quality that gets to the very essence of orchards.

The Ravilious/Common Ground travelling exhibition helped to reawaken the nation's sense of what they were losing. The whole apple culture which had been slewed by supermarkets importing foreign apples at odd times of year. There was no longer any season for apples and with chillers they could keep foreign apples on hold for months on end. It was not a level playing field

In 1989 Common Ground also published a 'Manifesto for Trees' and 'Orchards – a Guide to Local Conservation'. But far more had to be done. It was like a guerrilla campaign on an intellectual and artistic level. A fight for English apples, hearts and minds. They worked with the South Bank Centre, the Tate and the Crafts Council. Another publication, *The Tree of Life*, was linked to an exhibition with the Hayward Gallery which toured the country. The director Joanna Drew's words encapsulated the ethos behind the exhibition. 'Common Ground's philosophy emphasizes the particularities for place and the need to value and care for our everyday environment. This is a sound basis for practical life and for imaginative awareness. To find the universal in the particular is an essential part of the artistic venture.' Apple trees were now centre stage.

Sue Clifford grew up in Nottinghamshire. Her background was geography, but she worked with architects, town planners and landscape designers. She also lectured at UCL and the Polytechnic of Central London. 'Environmentalism before it was called that.' Angela King was brought up in Branksome Park, near Poole in Dorset. She worked in dress design at the sharp end in New York, but became disillusioned with the sleazy fashion world and campaigned to stop the fur trade. She returned to London, joined Friends of the Earth and ran their campaigns to save whales and ban otter hunting. Angela and Sue met at FOE and have known each other since 1971.

Common Ground, unlike the National Trust, did not want members. Common Ground was essentially a novel idea, a state of mind rather than a company or corporation. They are not a business though it was their business to plant seeds in people's consciousness. Their ideas slowly evolved, migrated and transformed of their own accord. Orchards and cider have benefited no end.

First Apple Day

Everything came together for the first large scale Apple Day held in London – 21 October 1990. Covent Garden had been a fruit market since 1654. 'Part of the under-cover section was the old apple market. But the powers that be had forgotten all about apples. Fruit had gone to Nine Elms in 1974. We contacted many people and decided on a gathering of apple and orchard people.'

Apples had all the cultural and mythological connotations they needed, deeply embedded in people's consciousness. The background philosophy was to raise interest in orchards. Their first poster for Apple Day had 592 apple names written around it. They looked wonderful. I still have one pinned up in my garden shed.

Angela and Sue then started finding people with trees to sell and apples to show. They wanted cider, so they asked Nigel Stewart from Bridge Farm, East Chinnock, and Julian Temperley from Burrow Hill Cider. Julian had just founded the Somerset Cider Brandy Co. so he brought along a wonderful curved, swan-necked copper Calvados still. Just the ticket. The spirit of adventure.

As Covent Garden was only just round the corner from the opera house, they set up a tent two weeks beforehand with James Ravilious's photographs and a table full of apples that people could try. Hands-on marketing. 'We threw everything we could at this single event. Brogdale, the mother orchard in Kent, came along with a hundred different varieties. We ran an old-style media campaign. Radio 4, local radio, even the World Service. The twenty-first of October was nothing to do with Nelson and Trafalgar. It was a convenient Sunday and the twenty-first had connotations of equinox. A sense of seasons changing.' All the local shops put up posters. 'England expects every man and woman to do their duty and make the best possible . . . CIDER.'

On the day itself, they had over forty stalls. There were cider makers, wooden cider presses, cider brandy distillers, mobile French Calvados stills, apple juice, apple merchants, apple identifiers, WI stalls, bottles, jams, preserves, tarts, flans, jugglers, acrobats, apple flambé, baked apple, longest apple peel competitions, apple bobbing, poets, illustrators, musicians, booksellers, chefs, journalists, actors, opera singers, mistletoe. All present and correct. You felt that 'Pretty, witty Nell Gwyn' might appear at any moment.

Seeing apple juice pressed out in the heart of Covent Garden was for many people their first taste of cider making and rural life. No different from Giles Winterborne in *The Woodlanders*. People drifted around just absorbing the atmosphere, taking on board images and ideas by osmosis. The seeds of future apple days were already growing in people's minds.

Common Ground then had to persuade people to do their own apple days. DIY apple philosophy. 'The Countryside Commission was interested. We launched *The Apple Source Book* and *The Apple Games Book*. Michael Barry from BBC *Food & Drink* came to see us. A lot of people were setting up their own local apple groups. We got articles into *Country Life*, *Country Living*, women's magazines, journals. Wildlife trusts. Covering every angle. All the national papers . . .' There was no stopping them but to turn one event into a national event across the country was a different challenge. And one that in the end grew exponentially.

In 1991 there were fifty events. Often it just required one highly motivated person in each village or town. Apples were celebrated in the House of Commons. They got through a ton of English apples. Then Common Ground had the major job of persuading supermarkets to take a wider range of English apples. 'Sainsbury's had twenty different varieties. Safeway were the best. They had fifty-three varieties. That was down to Teresa Wickham. She co-founded the Women's Farming Union to rescue the UK top fruit industry and bring farmers closer to consumers.' The whole point was that the apples were sourced from England.

Curiously, conservationists weren't interested but food writers and food journalists were. 'They were very good indeed: writers like Sophie Grigson, Keith Floyd, Philippa Davenport, Rose Prince. Carol Trewin was fantastic – *BBC Food and Farming*, Elizabeth David, Prue Leith. Antonio Carluccio and the pomologist Joan Morgan. All well aware of how important taste, texture and looks were with apples and how local varieties had a character all of their own. Provenance and terroir. Identity rooted in the English landscape.'

A Community Orchards pamphlet was published and then a Common Ground Apple Day website which registered all the Apple Day events county by county. This was consulted by thousands every autumn. By 1994 there were 150 Apple Day events nationally and by 1999 more than 600 Apple Day events. Quite an achievement. Common Ground were given a Special Glenfiddich Food and Drink award. They were shortlisted for the first BBC Food Awards. Apple Day continues to thrive as well as being a catalyst for reviving local varieties of apples that would have otherwise disappeared. The National Trust have embraced it wholeheartedly. A great day out sampling apples and mulled cider, measuring the longest apple peel and apple bobbing. Kids love it.

10

TWENTY-FIRST CENTURY CIDER

Fast-forward twenty years and the cider landscape has changed out of all recognition.The twenty-first century has been a very dynamic one for cider. A new generation has discovered cider country and cider is the new drink of choice, particularly for young women. High-quality cider is the norm not just in rural England but in towns and cities. People's eyes have opened to the possibilities. Bottle-fermented cider, *pét nat* and keeved cider are now popular. I have seen palette loads of cider being sent to an Irish theme pub in Russia . . . Cider in Japan, America, Canada, Australia and New Zealand is on the up. High-quality cider has gone global.

Something has obviously happened in the shires. Almost every village in England now has its apple enthusiast and community orchard. The new breed of cider makers come from all walks of life. Barristers, scientists, teachers, artists, potters, agricultural engineers, farm workers, thatchers, civil servants, journalists, publicans, musicians, rock band roadies, tree surgeons, builders, celebrity chefs, the list is endless. And their enthusiasm knows no bounds. A bit like the renaissance in cider making during the seventeenth century. So I decided to make a journey through South West England and visit some key apple and cider hotspots just to find out what was happening.

Some villages have cider clubs and cider festivals. These are the hubs of the village where the village is small and has no shop, no post office, no school and no pub. The social vacuum has often been filled by apples. An ingenious way of bringing the village together once a month and in the autumn to make the cider and then in summer to drink it. What could be simpler or more wonderful?

In the old days cider clubs were made up of old farm workers who had a cider press tucked in a shed alongside a pub or in a pound house. Most of these men had a desperate thirst and a keen interest in orchards. Cider clubs in Dorset and Somerset almost died out. But not quite. But it's a miracle any survived.

It is obvious that the social side is a key ingredient. Cider clubs are not new in West Dorset. The Chideock cider club has been going since the 1960s. Another neighbouring village called Monkton Wyld has three cider clubs. And that is a bit wild and hillbilly. This is how the cider knowledge and philosophy is kept alive. Underground, samizdat cider. Old sofas and armchairs come in handy next to the barrels. The 'chair of cider philosophy' is always available. Some still drink out of horn cups.

But these clubs are not to be confused with cider houses

like the Monkey House at Woodmancote in Worcestershire, not too far from Pershore. Here cider is kept behind a stable door and served in jugs, which is how all cider and beer used to be served. It has been in Jill Collins's family for over 100 years. In Somerset there used to be a cider house at Nempnett Thrubwell, near Bristol Airport, run by two old ladies. You just had to know which door to knock on. In those days pubs were often just people's front rooms. There was another undercover establishment tucked away in a farm near Ilminster where old men crept in by the back door. The farmhouse was so covered in ivy it looked as if it was abandoned. Good cover. I used to shear sheep there. Each cider house or cider club is a law unto itself. Times haven't changed that much. Elis - the Rose & Crown near Langport in Somerset still has a tap room with worn flagstones and no bar.

Cider houses may have dwindled but cider clubs have blossomed. One cider club I know well is at West Milton, near Bridport in West Dorset. Great, green rolling hills, flocks of sheep and small, thatched cottages. This cider club used to meet in an old barn but now they have a new clubhouse at West Milton Mill which looks like a small cricket pavilion. Instead of the cricket pitch there is an orchard. The orchard keeper is Margaret, the owner of West Milton Mill which was once the home of the naturalist and broadcaster Kenneth Allsop.

On Sunday mornings this club is a great spot. A bit like a Quaker meeting where people drift in and out. They bring food and share. Cider is on the altar, a wooden table set up in the middle. A good place to see old friends and chat. Communion without a sermon. Again, the ancient Greeks would have approved. Cider philosophy in action.

The West Milton club is twenty years old and they meet on the first Sunday of every month. When I visit a glass of cider is promptly thrust into my hand by Nick Poole, the aficionado and driving force behind the cider making. He also runs the local cider festival and is a builder, songwriter and banjo player. We test the first cider of the year. Early, light and very good. A lot of well-dressed women turn up in wellies. Not just a big boys' drinking den.

To find out what makes the cider club tick I make discreet enquiries. Sally, a local sculptor, regards the cider club as the heart of the village. 'It is very casual. Everybody helps to make cider in the autumn. When people ask what do you do in the village? I say we make cider. We have apple days, carols at Christmas. Wassail and even the odd wedding. Nerissa the local vicar used to time her services so that she could get to the cider club. A founder member no less. The philosophy of each cider club evolves in its own way. People can come here and talk about anything, which is what a community should be . . . The apples now hold us together.'

Then there is Miranda, a garden designer and psychotherapist. She regards the cider club as a wonderful coming together of people. Just very convivial, very grounded, down to earth and timeless. For Margaret, the owner of the Mill, the key thing is 'Enjoying life. The real magic is making apple juice and then cider. We had some very good results with last year's early cider: a blend of Morgan Sweet, Lady's Finger and Tom Putt.'

Having ideas is one thing, putting them into practice is something else. It needs a leading light behind the scenes. Nick Poole has not only rejuvenated cider making in the valley, he has spent ten years researching Dorset cider apples with Liz Copas. Nick is a bit of a legend in these parts. News of his intrepid enthusiasm for cider reached the Bath & West Show where he was awarded a gold medal. His keeved cider won Supreme Champion that year. Not bad for a local builder with a background in farming and catering.

Nick lives in Pear Tree Cottage, named after Pear Tree Farm. 'Not many pears here today,' he quips though he has made some excellent perry. 'I have lived in this parish all my life and they all die of old age.' I'm not sure if he is talking about the cider trees, the sheep or the inhabitants. I decide that he is talking about all three.

Nick is a local lad and by the time he was at school in the 1960s cider making was on its last legs. The estate at Nettlecombe still made cider and there was a communal press down the lane. About half a dozen farm workers would make the cider. In those days two cider farms in the next village supplied the pubs of Bridport and kept the rope makers happy. Most of the farm buildings are now houses and holiday cottages.

Forty years later Nick Poole tracked down one of those farm labourers called Jim and learned about cider apples. This is the vital bit of the equation, finding someone who knows his cider apples. All too often continuity has been broken and apple trees grubbed out.

It is curious how one thing leads to another. It was

horses that led Nick to cider. His wife Dawn rented a field, but it had apple trees on it. Horses cannot eat too many apples as they get colic. So they had to pick the apples by hand and, having picked them up, they decided to make cider. The rest is cider history.

Jim identified them as 'Dabinetts, Chisel Jersey, Bulmer's Norman, a Bramley or two and some Yarlington Mill.' Whoever had planted them had probably been to Long Ashton. Horses therefore kick-started Nick's interest in cider and opened up a world of discovery. As Nick says, 'I knew it was a simple process. The first year was 2000. Morgan Sweets and Bulmer's Norman. Wow. This is amazing. All wild yeasts. It led me on to reading all about keeved cider. I wanted to produce a naturally sweet cider without sweetening it. A bit of a purist. The holy grail was to make keeved, natural sweet cider, maintaining the fruitiness, aka the Normandy method.'

But Nick was adventurous. He met a French cider maker called Michel Ameline and went round cider farms with him in Normandy. Michel lives near Fécamp in the Pays de Caux. He makes *cidre bouché*, i.e. bottle-conditioned cider with the cork wired down.

The first time Nick made *cidre bouché* he went over to France and entered a competition. Amazingly, he won first prize. 'It took them by surprise. It took me by surprise. To be honest I had only made about four dozen bottles. Just grabbed three random ones and chucked them in the back of the car. That whole batch was perfect. Maybe the journey across the Channel helped it. Beginner's luck.'

Powerstock Cider Festival

Cider making is one thing, but starting a cider festival is something else. Cider festivals are wonderful events. They either make money for the cider makers and local publicans or for charity, or all three. West Milton and Powerstock are linked at the hip. But for a festival first make the cider. Sounds basic? It is basic. Nick Poole just went round all his neighbours who each put in £20 and then they had £400 to buy some second-hand cider equipment from a retiring cider maker in Somerset. They then built their own press. 'One of the members, Spanner Ward, was an engineer and had access to welding facilities. Everybody chipped in.' That was the start . . .

The festival evolved out of the cider club. There is a kind of cider logic to that. You make cider, share it around, have a good time and raise money for charity. Grass-roots thinking. The Powerstock Cider Festival is now famous up and down the country. It grew of its own accord. In 2000, after the first cider making, Nick had a tasting evening at the cider shed and fifty people squeezed in. He was amazed so many people were fascinated and enjoyed it. The following year he took over as chairman of the village hall, which had just been rebuilt. They didn't have a penny in the bank. So Nick thought he would repeat the cider-tasting evening. That was 2001. Not many were making cider in Dorset then. It had almost died out. Without quite realizing it, they were filling a cider vacuum. Not just in Dorset but in other parts of England.

Word went around. Nick never advertised and suddenly

people came out of the woodwork from all over the country. Hereford, Oxfordshire, South Wales, Gloucestershire, Yorkshire, even from Scotland. Then Michel came over from France. Word of mouth. Word rippled out. Cider makers brought their own cider, gave tasters and talked to anyone and everyone. Nobody had ever really done that before. The incredible, underground, samizdat cider-making network had surpassed itself. It was like a hedgerow cider university seminar with tastings.

The key to the success was in the organization. When 500 or 1,000 people turn up to a small village hall in West Dorset you have to be prepared for anything.

Nick told me how it worked. 'The system was brilliant. People bought tickets. Just like raffle tickets. A strip of five for £2. Pink, green and blue. No money changed hands inside the hut. First few years we didn't have much music, so people just talked about cider and asked questions. We had up to thirty cider makers. Lots of different tastes and flavours. Everything from vinegar to award-winning, top-class cider. All under one roof and if you didn't like it you could spit it out. We did have buckets.'

It moved from November to April. Then it got much bigger and they also introduced a cider competition. Powerstock has a real party atmosphere. 'People went home saying this was the best party in Dorset. They also raised lots of money for charity. Cider makers donated their cider.' I went to quite a few and really enjoyed it. I even took my father along. That was the year Hugh Fearnley-Whittingstall came along with a film crew.

But the cider festival is not just for old fogies, as Nick

explains: 'For the first few years local kids kept away. They wouldn't be seen dead in the hall with their parents. Then they realized it could be a lot of fun and cider was cheap – fifty pence for a third of a pint. When my own children and other children in the parish realized something wonderful was going on, word spread. They started coming and we had to hire bouncers. Then we did tickets in advance. It was too big. I think we peaked at around about a thousand queuing all the way across the school yard down the road. It was frightening.'

But having such large crowds has its own problems. Risk assessments – health and safety. That sort of thing. They consulted the police. Nick's brother was a policeman, but he steered well clear. Strangely, the crime rate in Bridport always went down to zero on Powerstock cider nights. Getting there youngsters crammed into taxis and when it was time to go back home they found their mobiles didn't work and neither did the phone box. The hills are very steep. So the kids either walked all the way back into Bridport or slept in people's gardens. 'We did once find someone asleep under a wheelbarrow. People would find odd bits of garment in their garden. Some even slept in the flowerbeds . . .'

These cider clubs are a throwback to the past but with a new dynamic. It must be to do with the archaic nature of the beast. Fruits of the earth, orchards and old, pre-Christian festivals. A strong social statement about who you are and where you live and who your friends are. Social identity, drinking good cider and having a good time. West Dorset is alive with cider. Nick reckons every village should have a cider club. The philosophy is to enjoy yourself.

Amazingly there is a second cider club in Powerstock up the valley at Nettlecombe run by Victor Crutchley. They use the beautiful old estate cider press and meet once a month on the full moon. Gentlemen of the press.

Apple identification

Where does this new interest spring from? For many it starts with a sense of curiosity. A need to know what something is and then to name it, whether rare flowers, butterflies or apples. It is all part of being English. We are a nation of gardeners and shopkeepers. One grows the apples, the other sells them.

Apple identification is an important part of the process and a key to the renaissance. Apple experts are called pomologists and they work overtime on apple days up and down the country identifying apples. No easy task. People stand patiently in long queues clutching apples in brown paper bags as if it was wartime rationing. A bit like waiting for the oracle at Delphi or reading someone's palm. People's interest is undimmed and such events are very special. They help to regenerate the knowledge that was lost over the last century or so. People's enthusiasm for apples knows no bounds. Tree nurseries have done very well indeed.

Many counties now have their own cider apple experts. Mother orchards are now being planted, living encyclopedias, county by county. The main thing is to make sure the name tallies survive, otherwise you are back to square one. The other thing is that identification can be a dark art. So DNA test results on apples are increasingly important to iron out inconsistencies and oddities.

In the Tamar Valley in Cornwall, Mary Martin and James Evans have spent a lifetime researching Cornish apples and have produced their own *Cornish Pomona*. In Somerset the redoubtable pomologist Liz Copas has written two books: *Somerset Pomona* and *Cider Apples*.

One county that was a 'black hole' because so little was known about its cider apples was Dorset. The cider makers had all but died out so there was often no one to consult. Back in 2006 Nick Poole sent apples into Long Ashton, which had officially closed in 2003, but Ray Williams was still working there. Ray passed them on to Liz Copas. Within a year or two Nick and Liz had started a project called DATA – Dorset Apple Tree Analysis – and went looking for old orchards. A classic piece of cider apple research work. Nothing beats meeting the apple in the wild.

Mr Pickford from Long Ashton went down to Dorset in 1938 and gathered as many cider apples as he could. It was very formal in those days. No Christian names. But there were trial orchards in Loders and Up Loders. Liz Copas found Mr Pickford's early notes invaluable as they listed cider makers' apple varieties. So they knew exactly where to look to start their adventure. Not quite like Kazakhstan but nor far off it either. Some of the orchards were very wild indeed with trees keeling over or smothered in brambles. You needed a machete for bush wacking. Later on I discovered Mr Pickford's full name was Phillips Thornley Hyde Pickford, which sounds a bit like a portrait artist.

Slowly Nick and Liz pieced the evidence together. They were real detectives. There were very few commercial cider farms except at Netherbury and Melplash. Much of central Dorset is chalk so the orchards are often on the edges or valley bottoms. They travelled far, right up into the Blackmore Vale around Childe Okeford and Sturminster Newton, as far as Shaftesbury. Then again at Leigh and Chetnole, near Sherborne. But not all farmers knew what they had. One farmer who had been making 10,000 gallons and importing apples from France had no idea. What amazed Nick was how quickly valuable knowledge was lost. 'Three generations sat round that table. Even the grandfather couldn't remember what they used.'

Over five years Nick and Liz narrowed it down to twenty-two Dorset varieties, which they propagated. Quite a few of them were 'new'. In 2020 nineteen varieties cropped at the same time and so Nick made single-variety ciders out of each of them. There was even a tasting day and thirty people turned up. A wonderful occasion and the first time some ciders had ever been tasted on their own and their qualities appreciated. We were all sitting at long tables in Rupert Best's barn in Hincknowle, Melplash. It felt historic. New horizons. Rupert's swan song as it turned out.

Liz had an equally down-to-earth approach. 'Not many with high tannin. Out of nineteen this year we got some well-balanced ones as single varieties. Nothing to replace

Kingston Black or Dabinett but a good Dorset cider blend: Golden Ball, Marlpits Late and Winter Stubbard.'

In fact, Nick and Liz continued what Long Ashton started 100 years ago. Now, after fifteen years of research, they have a good idea of what is out there. But to be of any real use the apple trees have to grow well, crop well and be disease-resistant. And, crucially, they have to make good cider. All those elements need to come together to make a successful commercial cider apple tree. 'Most are best blended. Golden Ball is fantastic. The tree grows well, crops well, a bittersharp – a good single variety.'

But you need several generations to establish local orchards. As Nick says, 'I have only been doing it twenty years and even if I had planted an orchard twenty years ago I would have made mistakes. I love Sweet Alford, a Devon variety. Knowledge is always held in families. Brings it home to you when you realize how long it takes to make a good blend.'

Back in 1938 when Mr Pickford came down from Long Ashton his list was very useful and Liz Copas used it for reference. In Hubert Warren's orchard in Netherbury he found Buttery Door, Best Bearer, Golden Ball, Mount Seedling and Woodbine from Perhay. Then from elsewhere: Crimson King, Ironsides, Smith's Sweet, Somerset Crab, String Pippin, Chetnole, Honeycombe, Improved Pound. I loved these names. Most were unusual in the wider cider world.

Seventy years later many apples had lost their names. So Nick Poole and Liz Copas painstakingly compared apples from different villages. Some had crept over from

Devon or Somerset, but others had no name. And they had to be given names. Which is exciting. Some new, some old. DNA analysis helped a lot.

Names like Sweet Sheep's Nose or Bell, Dewbit – Sour Cadbury, Sweet Alford, Reinette Obry, Sweet Coppin, Slack ma Girdle, Fair Maid of Devon, King's Favourite, Tom Putt, Dorset Longstem, Golly Knapp, Moonshines, Tom Legg, Stubbard, Marnhull Mill, Tangy, Dashayes Crab, Golden Ball, Charlie's Seedling, Golden Bittersweet, Marlpit's Late, Winter Stubbard, Puddletown, Filibarrel, Northwood, Dashey's Red, Cap of Liberty, Hunter's Ground, Joannies, Marnhull Bitters. Take your pick. Almost every village had its own variety which it clung onto.

Dorset is no longer a 'black hole', but there is a lot of work to do with propagation and blending to get a distinctive Dorset cider. It is hard work but very rewarding. Tasting the landscape. Cider that would have once gone out to Newfoundland on the cod fishing boats. Back in 1793 John Claridge reckoned that Dorset had 10,000 acres of orchard. So there is a little way to go.

Apple detectives

Dorset was only one of many recent apple research projects in the West Country. The apple and cider renaissance was widespread and spontaneous. So I went to visit some of these apple experts. Fruit growing was often in their blood. In the Tamar Valley, near Cotehele House, Mary Martin and James Evans have spent many years looking out for Cornish apples, pears and cherries. Mary's family were involved with horticulture, farming and salmon fishing for generations. I visited them for tea. Mary chatted away. 'Grandfather was the last miller at Cotehele Mill. Father was a smallholder of about eighty acres and lots of relations were involved in growing cherries and apples.'

Mary remembers a row of cherry trees her grandfather cut down for firewood in the 1950s. 'There were so many. End of the cherries. Eighty people were involved in the market slopes between here and Cotehele Quay. A mile. All south-facing slopes. Much of the fruit ended up in Covent Garden. Strawberries were sent to Glasgow and during the First War apples were sent to the front.'

Mary's Uncle Ernie was the fruit grower. Mary remembered him very well and when it was cherry time his job was to keep the flocks of marauding starlings from damaging the cherries. 'He had big gun at four a.m. It was really hard work. The women with big old trees went up forty-bar ladders. Very dainty in long skirts. Dangerous work if you fell.' These women were very intrepid and used to sing to each other across the valley. A bit like Songs of the Auvergne.

James and Mary have produced two books: *Burcombes, Queenies and Colloggetts*, about Tamar apples and cherries written by Mary's sister Virginia Spiers with Mary's illustrations; the second, with James, called *Cornish Pomona*. A very useful handbook that slips nicely into your coat pocket. They always put up a wonderful exhibition at Cotehele House on Apple Day weekend.

As we sip tea James and Mary are full of apple stories. '*Breadfruit* was a single tree that can be dated back to 1900 at Rezare near where James grew up.' DNA testing threw up some interesting parallels. It was the same as *Bloody Butcher*, which crops up in Ireland in 1951. 'Then there is *Pendragon*, named after the house where they discovered it. The original owner was Canon Andrews, vicar of Stoke Climsland and a renowned plantsman. It has a red skin which reaches deep into the apple, a right red blush. Incredibly healthy according to the experts.'

Listening to them chatting was music to my ears. Other key apples came from Cornish cider farms I had visited in the past such as Haye Farm at Lerryn, near Lostwithiel. Cider had been made there since the thirteenth century. It was a good location up the Fowey Estuary opposite Golant. Just down the valley is the site of St Cadix Priory, linked to Montacute Priory in Somerset. The monks always pick the best places. There are about forty different apples there and a quarter of them are unique so they are simply called Haye Farm 1–10.

The same thing happened at Felldownhead, near Milton Abbot, Horace Lancaster's old cider farm. This was the farm I'd visit as a boy on a bicycle. It was Horace who always used to say, 'Bread is the Staff of Life but Cider is Life itself.' The farm is now run by Vernon Shutler and his wife Theresa. Horace had his family farm over the Tamar at Lawhitton, so at least seven apples were named after that village. Lawhitton 1–7. Horace liked No. 7. He also liked Dawe No. 1, which he called the Rolls-Royce of cider apples, and Dawe No. 2, which was Ellis Bitter, an old East Devon variety from Honiton. Then there was Hockings Green raised by Mr Hocking of Coad's Green, Callington, in about 1860. And the famous Collogett Pippin, an enormous apple from Botus Fleming. The list goes on and on. Another Cornish world yet to be explored.

We have another cup of tea and tuck into the fruit cake. But what did the cider taste like in the old days? James reckons they had more tannin 200 years ago. Now generally very mild bitter. 'If you go back to some very old trees, even the pigs won't eat them. Two hundred years ago they needed bittersweets for sea voyages. Higher tannins.'

James and Mary have several mother orchards of their own. Then, about ten years ago, they helped Chris Groves, the orchard manager at Cotehele House, to plant a large Cornish mother orchard of 250 trees for the National Trust. Chris Groves comes from Bridport. His father runs a nursery and he is the fifth generation of nurserymen.

Getting the apple names right and planting healthy trees was very important. Chris plotted the orchard out with binder twine. In the future people can have graft-wood so they can propagate their own orchards. Rather like Ralph Austen in Oxford with his library of trees. Each orchard is a book that can be consulted. At Cotehele they make cider at Apple Day on an old wind-lass press.

Fig. 10.1 'The Big Squeeze - building the big cheese', Apple Day, Cotehele House, Cornwall, 2019

In North Devon much the same has been achieved by Michael Gee with Orchards Live in helping to get apple knowledge embedded into the community. Michael has written a very comprehensive book, *Devon Orchards*. Training courses have been run to teach pruning and propagation. Vital knowledge passed on. Good volunteers are crucial and a great asset. Luckily RHS Rosemoor at Great Torrington have planted their own North Devon mother orchard with forty-five rare varieties. And in South Devon there is an equally vibrant organization called Orchard Link. With storms and climate change round the corner, who knows what may happen with pests and disease?

Fig. 10.2 'Cidermaking at Westcott Farm', near Callington, Cornwall. Oil painting by Mary Martin, 1981

In Somerset most cider varieties are very well known and propagated all over the country so that there is no real need for a mother orchard. Though I did find one unusual orchard in Baltonsborough. Some varieties I did not recognize. It was a Long Ashton experimental orchard planted in the 1950s with Hereford varieties to see how well they grew in Somerset. In fact, every cider orchard in Somerset is a mother orchard laden with Dabinett, Kingston Black, Harry Master's Jersey and Yarlington Mill plus all the others in the band.

North of Bristol lies Gloucestershire which over the last fifty years has lost about about 75 per cent of its orchards. The decline is serious. But for the past twenty years the Gloucester Orchard Trust has tried to conserve the genetic characteristics of the 100-plus apple varieties that are unique to Gloucestershire, as well as the 123-plus varieties of perry pear. One of the leading lights is Ann Smith. In 2015 the Trust acquired a fine perry orchard at Longney, alongside the River Severn.

Then there is the Orchard Centre at Hartpury, near Newent. Thirty acres of orchards and wetland as well as the National Perry Pear Collection. The centre is run by Jim Chapman, a local solicitor who inherited three or four fields from the family farm. He is passionate about orchards and perry pears and with the help of Charles Martell they have laid the foundations of what is now a vital resource. Charles makes Stinking Bishop cheese, which is rind-washed in perry made from the Stinking Bishop perry pear, a soft cheese which does indeed stink. He also uses cider to make Slack ma Girdle rind-wash cheese.

Charles Martell is very much a driving force, being a perry pear man, cider maker, cheese maker and author. He has written books on Gloucestershire apples, pears and plums and is a distiller. Back in 1998 he started planting perry pears at the Three Counties Showground in Malvern and soon ran out of space. Two more orchards were planted at Hartpury and this then became the National Centre. A building was erected on the site to facilitate meetings. I visited back in 2008 and could see that it had enormous potential.

They now have 250 trees of 142 varieties. It is here that Peter Mitchell used to run his cider-making courses, which he has also given in America, Canada, Australia and New Zealand, helping many small makers get to grips with the chemistry of cider making. Peter was one of the cider makers who joined Andrew Lea on his trip to Kazakhstan. Passing on information about orchards, cider and perry is vital. Otherwise it will all be lost. Again.

In the Welsh Borders alongside Herefordshire the Marcher Apple Network has been identifying and preserving apple varieties since 1993. Over fifty varieties have been rediscovered that were thought to be extinct. They have five museum orchards, three in Powys and two in Herefordshire with over 900 trees.

But the largest mother orchard by far is in Kent at Brogdale, Faversham, only a mile or two from where Richard Harris's mother orchard was in Henry VIII's time. They have over 2,200 varieties of apple, 500 of pear, 350 of plums and 322 varieties of cherry. In the spring it is the most extraordinary sight. Acres and acres of blossom. Mother orchards are very important. You never know when the genes may come in handy. It is amazing how easy it is to lose a variety if a storm takes a tree or the label falls off or a bypass ploughs through an orchard. Or someone dies before passing on their knowledge or writing it down. Apple detectives are vital.

Perry Maison

No discussion of cider in Herefordshire, Gloucestershire or Worcestershire is ever complete without perry, the subtle and elusive half-sister of cider. Perry pears are extraordinary things. Small, hard and difficult to find in the long grass. They don't float, which means they are difficult to handle. They rot from the inside out and go off very quickly before you realize it. And with global warming and warmer autumns they drop off the trees about a fortnight early. But for the indefatigable and dedicated producers they do produce that subtle, beautiful drink that is perry, often naturally sweet with soft tannins. Perry is a truly exquisite drink on a summer's day almost elderflowery and far more elegant than most champagne. Perseverance pays off.

But perry pear trees are anything but small. In spring, in full May blossom at over 50 feet high, a line of perry pears can be magnificent. A sight you will never forget. Tall, elegant and long-lived. In the late 1990s on a sunlit May morning I was in Gloucestershire to see the Severn bore and after it had passed through I went to see the cider maker, salmon netsman and water bailiff Jasper Ely near Framilode. Jasper was a real character, large, bluff and bearded. He had worked on Severn barges and as a merchant seamen he had sailed round Cape Horn. There wasn't much he hadn't seen or done. He kept a cider orchard laden with a few Gloucester Old Spot pigs which made excellent bacon. Sometimes he went about with his shirt off. Covered in tattoos, he had a sailing ship on his chest.

What fascinated me that morning was not just Jasper and his tattoos but the sight of about half a dozen tall perry pear trees maybe 50 foot high on the other side of the river in full blossom. Their reflection rippled on the

Severn which was already topping its banks after the Severn bore had passed through, the 6-foot tidal wave carrying all before it. And then I saw the true majesty of these extraordinary trees . . .

Like brides bedecked in muslin swaying beside the river,
their lace reflected on the flooding tide,
tall ghosts that shimmer upon the river's vapour.

A lasting dalliance that blossoms and beguiles,
sweet tonic that makes the water shine more deeply,
sparkle in its tidal dance with petals on its surface

Tall trees that richly primed, make mirror of one's soul
Words that cannot yet reach, the subtle truth that nature
Beckons both past and future in its timeless meditation.

River of blossom that has its own currency. From then on I was captivated by perry pear trees and sought out perry wherever I could find it. The number of top-notch perry makers in England can be counted on the fingers of one, possibly two, hands these days. Outstanding perry is very rare. Tom Oliver, the renowned perry maker of Ocle Pychard, east of Hereford, often repeats the old adage: 'Cider is a hard taskmaster but perry is a fickle mistress.' Tom has deep respect for perry pears but agrees they can be problematic. They can turn overnight. Pressing them isn't always easy. They have unusual names like Gin, Brandy, Blakeney Red, Winnal's Longdon, Flakey Bark, Thorn and Gregg's Pitt. Fermentations are quixotic and unpredictable. But if you are lucky you can make something truly superb with soft tannins: it is sorbitol, the unfermentable sugars which make it slightly sweet. Perry is one of the great hidden secrets and treasures of the English countryside. Few, I suspect, have ever tasted the real thing.

Tom Oliver's grandfather had orchards and sold apples into Bulmers but then the price dropped in the 1920s so they switched to hops. Tom's apprenticeship was served at Seale Hayne Agricultural College in Devon and, like one or two other well known cidermakers, he graduated from the famous Cider Bar in Newton Abbot. Tom put it like this. 'What I had not been aware of, when I was drinking cider as a student, was how good it could be. What individual qualities pears and apples could bring to it. We were drinking blends dictated by the large cider makers, and blending is a key skill.'

It was while tasting perry made by another dedicated cider maker called Kevin Minchew from near Tewksbury that the penny dropped. The taste of his single varieties was extraordinary. Tom got really excited by it. Kevin was obsessed. 'He had an extraordinary skill and single-mindedness about cider making. He was bitten by it. You could go round and taste his various ciders and perries. I wasn't aware of how different their juices were. It was a voyage of discovery.' Tom started making cider and perry in the 1990s and has never looked back.

His philosophy is mostly common sense: 'In my mind it is a minimum intervention which involves harvesting the right fruit in the right condition, wild yeasts, spontaneous fermentation, no sulphur. Letting the fruit do the talking, then careful blending in the finished perry. A full expression of what that pear is capable of, just as nature intended.' Each perry pear tree is sacred. Each season a new voyage into the unknown. Tom has great respect even for the Blakeney Red which he calls the Ford Fiesta of the perry world. Good cropper and reliable.

The perry world is small but very special. Everybody knows everybody else. There is great respect for each other. Knowledge is passed on from person to person and rarely if ever written down. A sort of secret code. I was keen to work out the lineages of perry makers. To find out who had started this new wave of perry making. It was a bit like a jigsaw. One bit here, another bit there and then after a few days you can see the picture emerging

After visiting Tom I went a few miles south to see another perry maker – James Marsden of Gregg's Pitt, Much Marcle. Like Tom, James had been making cider for about twenty years. James lived up a long track in a small but very neat red-brick cottage which dated back to about the 1720s. James was busy labelling some perry. He told me that Kevin had learned, as he had, from a highly respected local lady called Jean Nowell. Jean lived at Lyne Down and it was she who kick-started the artisan perry revival back in the 1980s. I had met her a few times myself. She must have been in her seventies then. James Marsden was one of her biggest fans. Jean was the leading light in those days for perry makers, and James became her apprentice. He took his fruit up there and she gave her wisdom and learning in return for 30 per cent of the juice of the fruit. 'That is the way the deal worked and I continued to learn under her tutelage until I set up on my own.' It suited both of them. That is how you learn. She was James's guru, mentor and friend. Cider and perry philosophy in action.

James tells me some of the small but important things he learned. 'In so many ways Jean Nowell was extraordinary. She came to Lyne Down in 1984. Her approach was just so easy. She understood fruit variety. She taught me what I now pass on to others. Use all of your senses to determine when you are going to press. So I talk to people. How does the fruit feel? Is it waxy to the tongue? How does it smell? Does it smell of sweetness? Does it smell ripe? Cut it open with your penknife. Are the pears ripening from the inside? Are they bletting? Are the pips black? White? Or somewhere in between. And, finally, how did it taste? I learned from her how tannin confuses the palate. Some people confuse tannin with acidity and vice versa. So all those little nuances. These were the sorts of things we talked about.'

But where did Jean learn from? 'Jean Nowell learned a lot of it herself and she used to talk about a friend who was at Colwall. The people who owned that orchard. The other person Jean spoke about was Andrew Lea from Long Ashton. Andrew also gave me the confidence to start keeving fifteen years ago.' Her father also made cider.

The last time I saw Jean was at a very special tasting in James's orchard. A vertical tasting. Three different vintages of a single bottle-fermented Thorn – champagne method: 2003, 2013 and 2015. A very rare event under the very

Thorn tree. Kevin Minchew was there. Another of Jean's 'boys'. Sadly, Jean died about ten days later. But her perry knowledge lives on.

Spirit of cider

Back in the mid-1980s Bertram Bulmer realized his long-term dream of having a cider brandy distillery in the Museum of Cider in Hereford. After a bit of haggling Customs and Excise gave Bertram a small-scale museum/ educational licence. Calvados stills were imported and the distillery was named King Offa. This was the first time cider brandy had legally been distilled in this country for over 200 years. Quite an achievement.

Word of Bertram's success drifted down to Somerset and Julian Temperley of Burrow Hill Cider realized he could do the same thing but on a much larger scale. Julian nipped over to France and acquired several old mobile Calvados stills and started work in 1989. Over the years he brought quite a few stills back. The two main Calvados stills were christened 'Josephine' and 'Fifi'. They became the backbone of the Somerset Cider Brandy Co. More importantly, Julian Temperley acquired a full-blown commercial licence to distil cider from Customs and Excise. It soon became a thriving business. The spirit of cider matured in oak barrels. Cider brandy is now twenty years old. Equivalent to a malt whisky.

I went to Julian's farm and had a chat about distilling. As I drove to Kingsbury Episcopi, I caught glimpses of the tall elegant Cider Monument commissioned by Pitt the Elder. Burrow Hill is very distinctive: it has one tree on top, a sycamore. I turned into the farmyard. I had been here hundreds of times while cider making. It was teatime. The kettle was on the Aga and the 1763 'No Excise on Cyder' clock was ticking away in the corner, minding its own business, as it had done for over 250 years.

Julian came to Pass Vale Farm and Burrow Hill in the 1970s. He started out with 4½ acres of orchard and used the large wooden press to make cider. The previous owners, the Ducks, had been around for donkey's years. The soil is very good, on the borders of South Petherton; grade one land. Cider apples were excellent. According to Long Ashton, Kingsbury Episcopi is one of the three vintage cider areas in Somerset along with Baltonsborough and Wedmore. Fifty years have passed since Julian started making cider. He now has 180 acres of orchard, some of it at Over Stratton. He also buys cider apples in from Jamie Montgomery at North Cadbury. Excellent cheddar. Excellent cider apples.

Julian's family have been in Somerset for centuries. But there is a Cambridge connection as well. Julian's father, the 'Professor', who lived to 102, used to examine PhDs with Alan Turing at King's College, Cambridge. During the war the 'Professor' was an expert on magnetic mines and homing torpedoes. Very useful for sparkling cider and distilling.

'How did I get into cider?' Julian is in reflective mood. 'You fall into cider by mistake. Some are born into it. Others find themselves owning a farm that always made cider. In those days we made below the purchase tax threshold. Cider was taxed like sweets. You didn't pay alcohol tax if you sold less than about fifteen barrels a year.' Purchase tax was paid on luxuries. Was cider a luxury or a necessity? 'Depends where you were.' In the old days illicit cider brandy was called 'Necessity'.

When Julian started cider making almost every village had its cider maker. There were five cider makers surrounding Langport and two in Kingsbury. 'All gone in the last twenty years. Cider houses have become second homes. But there is now a new breed: a profusion of small cider makers with a high investment in time and equipment.'

Julian again waxes lyrical. 'Fifty years ago the cider maker was relatively better off than his customers. Now it is the other way round. The customers are better educated, better off and have bigger cars. They often come many miles, their age is older. They can afford anything they want. So the market has definitely changed and we have changed with it.' Cider philosophy has moved up a few gears. The debate is all about quality, not price. Instead of a race to the bottom it is a race to the top.

But Julian also reckons that cider has a slightly schizophrenic identity. It crossed class boundaries very easily. It was the fuel that kept the farm world going, but at the same time it also had pride of place on the gentry's dining table. Farm worker's cider and vintage bottled cider for the farmer's family and their guests. Dual-purpose cider. Cider is a broader church than anybody realizes.

Julian is proud of his old-style cider philosophy and sticks to his guns. 'There is nothing we do here now which is essentially newfangled. It is all in Worlidge's book – and *The Herefordshire Pomona* – the philosophy of what we do would be understood by both of those.' So time has almost stood still in this part of Somerset. The cider clock has deliberately been wound back.

As the longcase clock ticks happily away, Julian gets some glasses out and we try his latest experiment. He has been making bottle-fermented sparkling cider since 1990. Usually bone-dry. Stoke Red and Kingston Black. As Julian says, 'It is quite nice to be able to go to the shed and pick up something like this', as he brandishes another bottle of his new demi-sec sparkling bottle-fermented cider. As far as Julian is concerned, 'The cider world remains a largely forgotten jewel of the English food and drink crown . . .'

Fig 10.3 'Bottle fermented sparkling cider maturing in a *pupitre* or 'A' frame', Burrow Hill Cider Farm, Somerset, 2021

We then go to the distillery where I pay my respects to Josephine and Fifi – the brace of copper Calvados stills. I first made their acquaintance in about 1991. Elegant, robust ladies who came over from France on a cross-Channel ferry via Le Havre and Portsmouth. Shipped in from Normandy under a canvas tarpaulin. Seasoned and spirited, these ladies stand no nonsense. They work long hours and are very reliable. They came highly recommended with references. Both had given very good service to small farmers in the Pays d'Auge. When you work with them you have to treat them with great respect and pat them affectionately from time to time. Even polish them and talk to them.

In France Josephine and Fifi had been peripatetic Calvados stills mounted on the backs of lorries. A bit like barnstorming only this was from one cider barn to another, distilling late at night when entire extended families would turn up to watch. I think each family member was allowed 5 litres of spirit, but they had to be present. So the same gang of cousins turned up night after night at different farms. Josephine was large and reliable and in her late forties. Fifi was smaller, and faster, but a little unpredictable. Fifi had miraculously survived the war and the German occupation. Fifi was named after the French au pair on Julian's farm at the time I worked there. She used to bring out our coffee at tea break. Even hot potatoes if we were on night shift. And then she would have a sneaky roll-up with the boys.

Both stills were '*Fabriqué à Paris*', 'Made in Paris', by E. Gazagne 6 Boulevard Richard-Lenoir 6 Arrondisement. These were continuous stills thanks to the ingenious engineering of Aeneas Coffey (1780–1852), an Irish exciseman and entrepreneur. A third working Calvados still has recently been imported. She is called 'Isobel-Marie'.

Julian often talks in a roundabout way, but it is always

Fig. 10.4 'Julian Temperley in the distillery', Somerset Cider Brandy Company, 2020

fascinating. 'Worlidge mentions cider brandy making as a normal part of the cider world. That was the philosophy we wanted to follow. Distilling is a natural progression from cider making.' Julian was following in the footsteps of Robert Gibbes, *c.*1535, the distiller and prior of Montacute Priory only a few miles away. *Plus ça change*.

The bonded warehouse round the corner is impressive. Acres of barrels stacked two high. 'With distilling we prefer French oak over American oak. Limousin is bit cruder than Allier, which is tight grain. We had American white oak and French oak tasting with *eau de vie*. American tasted a bit blonde whereas the French oak was classy like Audrey Hepburn . . . There was a huge difference. We are going to have English oak barrels soon.'

Last year they distilled 140,000 gallons of cider. The

crop of a lifetime. Julian wishes to leave something for the future. 'It is quite an idea to distil something and leave it for twenty years. I will then be ninety-five years old, which is still short of my father's age.'

For Julian it is not just the quality of cider and cider brandy that is important, it is how you go about making it and maturing it. A way of life first. A business second. A dream, an idea: the orchards are far more than just the trees. 'A Norman trick this drink from apples. Distilled and fiery a hint of orchard on the tongue.'

Fig. 10.5 'Julian and Di Temperley in their bonded warehouse. Cider brandy well matured - oak matters', 2020

Women cider makers

Over the centuries many women worked in the fields. They were paid with cider just as the men were. In 1843 in Devon women were paid 7*d*. a day for apple picking with a quart of cider and 10*d*. a day at harvest plus cider. Farmers' wives also played an important role in cider making but they were often behind the scenes and their story untold. Making cider is one thing but selling cider and marketing it is something else. However, in the last twenty years an increasing number of women are now fully-fledged cider makers. It is a very broad church.

At the one end of the scale in Much Marcle is Helen Thomas at Westons. The company was started by her great-grandfather. Helen became managing director in 1996 and has been at the helm of the multimillion pound business for twenty-five years. At the other end of the same village is Helen Woodman who makes artisan cider with her partner James Marsden of Gregg's Pitt Cider. Her day job is with Worcestershire Wildlife Trust. Often women work on the cider presses just as long hours as men.

Today there is a group called Cider Women, which includes Fiona Matthews of Bartestree and Susanna Forbes of Little Pomona. Susanna started her career as a drinks journalist in London. Her husband worked in the wine trade. They live near Bromyard. Little Pomona was started in 2014 and now has a wide range of ciders. In 2019 Susanna, with Elizabeth Pimblett, helped organize a large cider banquet at the Green Dragon, Hereford. People came from all over the known cider world. Spain, Luxembourg, Poland, Japan, France, America, Norway, even Somerset. It was memorable. The Mayor of Hereford, Kath Hey, wore her gold chain of office. Very regal. Everyone was there including Gillian Bulmer, Bertram's daughter, who did a lot to help set up the museum as well as working in Bulmer's orchards.

In Somerset there are several families in which women have been involved with cider making for generations. Perry's Cider at Dowlish Wake has had at least three or four ladies bearing the name of Mrs Perry! All holding the fort behind the scenes. Then there is Jan Rich of Rich's Cider near Burnham-on-Sea who took over the reins from her father back in the 1990s. Jan is quite a dynamo. When she was young she would go out into the orchards with her father Gordon and estimate the crop on the ground – no easy matter with long grass – and then make the deal and shake hands on it in the orchard. Then, at Bradford-on-Tone, Louisa Sheppy of Sheppy's Cider runs an upmarket deli, restaurant and café, as well as selling cider and handling all the supermarket orders. At Burrow Hill and Somerset Cider Brandy, Julian and Di Temperley's daughter Matilda now has a major hand in running the business. Then there are young women cider makers like Beccy Leach of Wilding Cider, near Chew Magna, who used to run Birch restaurant in Bristol, and Isy Shultz of Barley Wood, Wrington.

Down in Dorset, near Dorchester, there is Penny Cake at Mill House Cider Museum at Owermoigne. They have a large collection of old cider presses and scratters, as well as hundreds of grandfather clocks and a vibrant nursery with ninety types of seed potato. Every autumn they make cider. Penny runs it with her mother and sister. At Monkton Wyld there is Sophie Burleigh and Jyoti Fernandes. Smallholding cider. Then there is Rose Grant at Winterbourne Houghton who had a long career in the air force and ended up as an avionics expert. A high-flyer. She used her scientific expertise to great effect with cider making. She was inspired by Andy Banwell, a local thatcher who caught the cider bug from my brother and myself when we made the cider in Durweston on a Victorian horse-drawn press in the early 1980s. Scratters at dawn. Cider wasn't bad either.

In Devon at Countryman Cider, near Milton Abbot, Therese Shutler works with her husband Vernon on the cider presses every autumn. Hard work in cold weather. Horace Lancaster's old place. Then, in the Blackdowns, Selena Mitford works as a cider maker for Alex Hill at Bollhayes. She is also a thatcher. And in South Devon at Loddiswell, near Kingsbridge, Heron Valley Cider is run by Natasha Bradley and her sisters. An all-women cider team. Natasha read psychology at Plymouth University before pursuing a career in marketing. Her philosophy is: keep it simple. Local apples – natural yeasts. Her parents were both cider makers. Her father was an actor. She sees cider as a key part of the modern food and drink movement, particularly when cider is paired with seafood. Her optimism is infectious.

Also in Devon, at Huxham Barns, a few miles north of Exeter, is a young and adventurous cider maker – Polly Hilton of Find & Foster – who with her husband Matt has made cider for five years. They are in the right area. Horrell's, well-known Devon cider makers, were down the road in Stoke Canon. They closed in 1974.

Last year Polly and Matt had a midsummer cider party to thank all those who had helped them pick up apples the previous autumn. It was a lovely evening and the light was just right. I had a chat with John Horrell who produced a bottle of cider from a shopping bag. 'Do you want some of this,' he said as he leaned against the barn door. Deep, dark and sweet. How could I refuse. Devon cider used to be medium sweet but complex. 'That was the last batch we ever made,' he said. Bottled in 1974. It was magnificent. Just a screw cap; nothing fancy. It's not often you drink cider forty-six years old.

Polly has her eye on those orchards. 'We are just starting to use fruit from an orchard in Rewe. A lot of the trees were grafted from the Horrell's orchard so there is continuity. Good range of fruit – a lot of sharps. We get so many bittersweets.' This year Polly found an old apple in Bradford Speke when all the other apples had been eaten by pheasants. This one remained rock-hard. No idea what it was. DNA will help . . .

It's excellent to see young people in their late twenties

Fig. 10.6 'Polly Hilton in old orchards', Upton Pyne, Devon, 2020

dive in the deep end, but I was keen to know how they started. Polly had been at Cirencester Agricultural College and had studied equine matters, plus wine making. She then worked for an animal nutrition company but wanted to leave. The catalyst was a study tour funded by the EU in Tuscany, which was about adding value to produce. She visited Slow Food producers and farms and then met someone who told her there were many neglected orchards in Herefordshire that he helped to look after. 'I then wondered if the same was true for Devon?'

'The first cider I made was 2015.' Matt, who is a carpenter and boat builder, helps her with the hard graft. Polly just wanted to start a small business. 'I read a Natural England publication that said how many orchards had been lost or neglected. Also got the information pack from the Trust for Endangered Species and went looking for orchards. We soon realized that nothing ever gets done until it becomes economically viable or sustainable. I then thought, I can look after these orchards and make cider in return for the apples. So many were going to waste. We were both brought up here. Strange as it may seem, I was not a dedicated cider drinker before this. We had to go to Tuscany to have the conversation . . .'

Polly then started talking to orchard owners. 'Everybody was so enthusiastic to see something being done with their apples. Walking into all these orchards was so amazing. Just incredible, the variety of apples. Not just visual but the taste. Walking round and tasting all these apples and thinking how much potential there was for Devon apples to make really good cider.' Without knowing it, they had picked a really good area. Prime orchard land since Domesday.

The breakthough came from an unusual source. They had a friend called Adrian Sargood who used to have an orchard near Christow – a medical physicist who made very good cider. I had known Adrian in Bristol forty years earlier and he was at their cider party. His sister ran the Bristol Packet in Bristol Docks when cider was still the order of the day for dockers and Dixie was selling cider from his thatched cider wagon.

Drinking Adrian's cider made a lasting impression on Polly and opened her eyes to how good cider can be.Then the orchards found them. Polly had a brainwave. If she saw a good orchard, she knocked on people's doors. If no one was in, she would put a letter through the letter box saying, 'Would you be up for us helping you with your orchard?' That is how they found the location for their cider works. 'The orchard hadn't been pruned for thirty years. It is now smartened up and bears good fruit.' They had landed on their feet.

But Polly did not stop there. She was very active in the 'learning department', finding out all she could about cider apples and orchard management. Loads of courses, even one on scything. 'Many were run by Orchards Live. Such a good charity. And Orchard Link in South Devon.'

The wine module at Cirencester was run by Susan McCraith MW. It set Polly up for life. 'Really supportive and very helpful.' With Matt she went to a cider maker in Ireland. Then they rowed down through the canals of France in a dingy to work on Simon Tyrrell's vineyard on the Rhône. They learned a great deal. Wine and cider science are almost the same.

Polly is philosophical. Her approach is one of utmost quality and common sense. 'Being inquisitive, we came at it from the right angle. Rather than the old farmers' OOH ARRH SCRUMPY approach. We found the apples by simply using our eyes and asking nicely. We stumbled upon them. Looking after an orchard takes time.' Mary Quicke, the dynamic cheddar cheese maker, has been a great support to Polly and acted as mentor. They are just up the road at Newton St Cyres. The name Find & Foster alludes to finding the cider apples and fostering the orchards. A brilliant piece of marketing. I get the bread and cheese out. Bottles are opened and the great tasting begins . . .

Secret orchard

One thing I learned from Polly was that chance plays an important part in cider making. A year or two ago, when I was judging cider at the Bath & West, I came across an unusual but excellent medium cider. Smooth, dark and complex. It won its class and then Reserve Champion. The cider was called simply 'Secret Orchard'.

It is always a joy to find young cider makers who know what they want to do and do it well. When I discovered that Secret Orchard was based at Nettlecombe Court I was even more interested. Yet another lost domain in west Somerset. An Elizabethan house and grounds. It had it all: carvings, plaster ceilings and overmantles, undermantles, large fireplaces, long tables, secret rooms, creaking staircases, the odd ghost and elegant Italianate stables with parkland and countless aged oaks. Demesne of the Trevelyans going back to Domesday, more or less.

As for oak trees, an eccentric friend of mine nicknamed 'Bins' has gathered acorns there for nearly thirty years. On his hands and knees like a squirrel, in all weathers. He picks about 2½ tons every year, then sends them up to nurseries in Scotland. Highly sought-after. High-quality, slow-growing oaks that produced fine timber for ship building in Elizabethan times. Saw off the Armada. Oak and cider go well together. Cider makers are a bit like acorns. The best ones grow slowly. Well preserved. Tannins, you see.

The Elizabethan house is now a field studies centre. The owner of the estate, artists and craftsmen all now live and work in the stables and outlying houses. Nettlecombe is now returning to its roots with top-notch cider making. Another journey beckoned, exploring cider country.

To reach the Secret Orchard I travelled across the Blackdowns to Wellington, where Julian's cider clock was made, then to Wiveliscombe, taking the Watchet road which winds through steep hills and on the way passes Combe Sydenham house, surrounded by forests. Elizabeth Sydenham married Sir Francis Drake. Sea dogs get everywhere. Next stop Monksilver and Nettlecombe Court. The long, curving drive led up to the house, 'a many gabled mansion of red sandstone', and a small church, all nestling in under the hills. Combe of Nettles.

At the Italianate stables, I meet Joe Heley, one half of

Secret Orchard Cider. Joe's barrels are sandwiched between stalls originally designed for hunters and pedigree stallions. He is up to his eyes in bottles and fermenting vessels. Vats of cider quietly bubble away. The other half of Secret Orchard is Todd Studley, a chef who lives in Falmouth. They were at school together in Minehead.

Joe has only been making cider for five years but, like Polly, he has learned the ropes very quickly. Serendipity also played a part in finding the orchard. We wander through the slightly ramshackle stable block which houses a menagerie of local businesses. Joe is a tree surgeon who studied horticulture and arboriculture and has pruned cider orchards, including Sheppy's. Joe's cider business started from a chance meeting. He had a notion that when he retired he would have a couple of apple trees and make a little cider. His story is interesting. 'I was saying this at a dinner party at Nettlecombe one night – a secret supper club that the blacksmith's wife Liz was doing. You turned up and you did not know anyone. You all sat at one big table. Tom Wolseley was there. His family own the estate and he said, "We have a load of apple trees no one's doing anything with in the walled orchard garden." They were planted according to a design they had found from a map of 1500. He also said, "There's some space in the stables and you can get yourselves going."' Joe met up with Tom the next day and grabbed the opportunity. The rest is history.

So the orchard fell into Joe's lap. 'I hadn't been planning it as such and I thought I had better go for it.' Very quickly he got Todd on board and applied for some Defra and EU rural grants which were a godsend. 'We bought a small basket press and off we went. A lot of work. We use two hydropresses now. Quick and easy.' They started selling cider in 2015. Got a silver at the Bath & West in 2016 and gold for the medium the Exmoor Mellow in 2017. That was the one I judged. A lucky find. Good orchards.

The Bath & West cider competition is probably the largest in the world with over 600 entries: 450 UK entries, 100 international and 50 newcomers. Seeing all the demi-johns and bottles lined up in the marquee is quite a sight. Every shade of golden brown, from straw to almost walnut. Judging takes two days. The geographical spread of prizes is very gratifying to see. Globally, cider is a force to be reckoned with at every level.

But what about the secret orchard? It was in a walled garden and was very overgrown. Chickens abounded. It had once been a fine garden and in 1842 the landscape architect John Claudius Loudon noted, 'There is an admirable kitchen-garden here, with the walls covered with the very best kinds of peaches, nectarines and pears, all in fine order, while the fig ripens as a standard. Even the Nettlecombe cabbage which Mr Elworthy had raised between the Paignton and Cornwall cabbages.'

So the walled garden was indeed a paradise – *pairi-daêza* – in Old Persian which is after all a garden walled around, with a microclimate all of its own. Good oaks, good acorns, good cabbages, good deer. Why not good cider?

No doubt it is down to wild yeasts as well. As Joe says, 'You can feel the history.' What really interests me is that there is a record of the apples grown at Nettlecombe Court, taken from John Trevelyan's common-place note book of 1582, which refers to Leather Coats (Royal Russetts), Red Star, Glass Apple, Essex Apple, Dorset Apple, Catsheads, Permains and Pippins and costard apples, as well as something called the *Domine Quo Vadis* and Pace's pear, named after Mr Pace of Brentmarsh who supplied the grafts and, of course, a fig tree. The Garden of Eden was complete even in 1582.

As to cider philosophy, Joe is down to earth. 'We bring on the flavours that are already there. We take a lot of pride in our blending and wild yeasts. Letting nature do its work. A lot of our thinking is traditional techniques on modern equipment with a bit of modern science. Then letting the cider speak for itself.' A fine aphorism, echoed by many.

The labels on the bottles are excellent: *femme fatale* and *fin de siècle*. Eighteen nineties. 'Always wanted it to be stylish and different. The way we market the cider is that the bottles say, "Welcome to our world and sense of Nettlecombe and where we make it". I like to think we are the real deal with the images we put out.'

'We tell people about the way we live, the real joy of making cider and the fun we have. Sadly, as we get busier, more time is spent on the business side. Quite a few restaurants have the cider. We sent a pallet load to Amsterdam last month and Tom Kerridge has it in one of his restaurants.' As to the cider apples, they have access to an orchard at Stogumber where they get a good crop of Harry Master's Jersey. HMJ as it is known, a fine cider apple. They are also trying keeved cider with Yarlington Mill.

Joe still keeps up the day job but now manages an estate with 300 acres of woodland. A lifestyle choice. A way of life. Cider and cider philosophy is obviously a lost domain that has yet to be rediscovered by many. Others never left it. Local apples, walled gardens, wild yeasts, good tannins and a hefty dose of mystique and history. As well as acorns, nettles and an obliging landlord. Secret orchards. Just the ticket. Crimson King, an early season cider, sells well down in Watchet. I am sure the Ancient Mariner would have liked a keg or two for his voyage down south.

New horizons

Since 2000 there has been a real renaissance in artisan cider making all round Britain. I have heard that there are as many as 650 cider makers in the UK at present, three-quarters of whom started up in the last ten years, which is brilliant. It is often a lifestyle choice, a second career and they are as keen as mustard. There must be budding cider makers in almost every county these days. But let's not forget the many well-respected, well-established cider makers who braved it in the lean years. In Somerset Roger Wilkins of Wedmore has at least two Banksy's in his cider house, one with sheep dropping by parachute. This is unreformed cider making, the same as it was in the 1950s, with a bunch of regular workers in overalls, all sitting down clutching a glass of cider and gnawing on a lump of cheese. Even the men on the recycling and refuse lorry time their coffee break to be at Roger's for 10.30 a.m. In Somerset, unlike other counties, farm cider never went out of fashion. It was always available. The party faithful still needed their gallon a day. That's 300 gallons a year per person, excluding Sundays. But there are far fewer farm workers these days. Most of them drive tractors that are worth the same as a small house.

Welsh cider is now almost as popular as Welsh rugby. And north of the border in Scotland, Jean Nowell's son Max makes cider in Dumfriesshire. Converting the Scots to dry cider is no easy matter. They sometimes need a long drink to counteract the effect of Scotch mist. Northern Ireland has several passionate cider makers in Armagh, and south of the border there has also been a fair bit of activity over the last ten years. I once gave a talk at Drogheda on the history of cider, a fantastic opportunity to meet Irish cider makers for which I was very grateful. Some real characters. It is not just what you make of cider; it is what cider makes of you. Some are also trying to find the elusive Cockagee. Cider reality is at long last catching up with Irish mythology.

Cider is also now a global phenomenon and that is brilliant. It means that cider is here to stay, and cider will revive itself and take root in many unexpected places. On my last trip to Herefordshire I met a Japanese woman called Yuki at Ross-on-Wye Cider who wanted to set up a cider bar in Tokyo, and an American called Matt who had studied English at Aberdeen University. He had just planted 3,000 cider apple trees in Ohio. Brave man. Then there was Esteban, a Mexican who had small cider farm an hour and a half south of Mexico City. At the autumn event at the Cider Museum we tested cider from Norway, Japan, Poland, Spain, Germany, Italy, Canada, as well as quince cider from Luxembourg. Ross-on-Wye had a palette load of cider going off to Russia and another to Belgium. Cider was crossing boundaries in more ways than one.

On the industrial front, I just hope that large cider makers do not lose sight of what really good artisan cider tastes like. It is to be hoped that the plethora of sweet fruit ciders is a passing phenomenon as young drinkers slowly find their way to the deeper, darker, more complex ciders with more tannin. If they don't they are missing out on so much. White cider has slowly, thank God, bitten the dust. These fruit ciders are like alcopops; they have an almost neutral apple base and are topped up with fruit flavouring, whether it be lemon, lime, cherry, strawberry, toffee apple, blackcurrant or raspberry. And, horror of horrors, rhubarb and custard. Maybe fruit ciders should be taxed differently? Sadly, whenever the British Government is in trouble it always tries to increase the cider tax . . . without defining cider or differentiating between artisan and industrial cider makers. It is a bitter-sweet relationship.

As far as international cider is concerned, there has been a marked resurgence in American cider. Official stats show around 1,000 makers in the US, with even one in Alaska, as opposed to around 650 in the UK. But the real figures are probably much higher as many small cider makers often do not register. I even heard the figure of 2,500 cider makers mentioned for the US. Mind you, they have 100 years to catch up since Prohibition, but they have the bit between the teeth and are making up for lost time. Many Americans also come over to the mother country to hone their skills and pick up tips. Also, their innovation in cider fermentation knows no bounds. Some use beer yeasts, wine yeasts and all sorts of mixtures. On skin, off skin, maceration, keeving. I expect there will be salami-flavoured cider before long. A cut above the rest? Old cider makers used to put beef in their cider to give the yeast something to feed on . . .

Personally, I prefer pure, unadulterated, dry cider with high tannin and deep, complex, fermented flavours. Some even have phenolics akin to malt whiskies. Cider should make you really appreciate where it comes from and the care taken to make it. Cider that makes you pause and think. Nearly always I like to get to know the cider makers as well. Some I have known for forty years and that is wonderful. With Alex Hill of Bollhayes Cider in the Blackdowns, near Clayhidon, we discovered that we were both in the same chemistry class at the same school. Understanding the chemistry of cider is a vital ingredient. Technique and science are as much an art as an intuition and instinctive skill. The rest is nature, soil, climate, the age of the trees and good luck. Alex makes superb very dry bottle-fermented sparkling cider and ages it. A rare treat.

These days you never know what is round the corner. One enterprising wine maker in Sussex, Ben Walgate, of Tillingham Wines, down near Rye in East Sussex, makes cider in *qvevris*, large Georgian clay amphora set in the ground. There are bulbous and very large, containing 250 or 500 gallons each. Skins, stalks and pips are sometimes included. Press the apples, pour the juice in, stand back and leave for several months, just as they did with wine 8,000 years ago. The cider has the same hints of lemon skin, stone fruit and honeyed beeswax. Is this the future? Certainly a new horizon. Another cider maker across the pond in Virginia also uses imported *qvevris* to make cider.

If cider keeps reinventing itself it means that the culture is alive and kicking. Quality will out. Ideas will distil themselves into projects and cider farms will blossom once more. It is a long way to Tipperary and it's an even longer way to Kazakhstan.

Conservation

Conservation is a major issue. It was only with Common Ground's intervention back in the 1980s that fruit trees, i.e. old cider and perry pear trees, were included in the list for Tree Preservation Orders. Up to that point they had no protection at all. Many local councils have no idea of their importance to wildlife or genetic diversity. Old, mature orchards with cider trees and ancient perry pear trees are like gnarled citizens, humble giants that keep fruiting year after year. These trees, known as veteran standards, are survivors. Preserve them at all costs. Each tree is its own microbiological city. Home to all sorts of other wildlife: woodpeckers, bugs and beasties, microbiological flora and fauna, fungi and wild yeasts. Let alone being visually stunning.

Orchards are multipurpose wildlife havens and should be treated as such. You can often graze sheep beneath the trees and I have often sheared sheep in orchards. Welcome shade. But orchards are often at the mercy of planners and vandals. For a number of years Gillian Bulmer's special orchard west of Hereford was under threat from a bypass, but there was a last-minute reprieve when the local council leaders were toppled. Then again, a famous wild pear tree, the 250-year-old Cubbington pear tree in Warwickshire, was felled in October 2020 to make way for HS2. Just to shave a few minutes off a journey that nobody really needs to make. Closer to home in Somerset there was blatant piece of perry pear vandalism near Yeovil in August 2020 where a youngish man, who shall remain nameless (but he should have known better), brought in cowboy tree surgeons on a Sunday morning and massacred a tall perry pear tree in his parent's garden. An Elizabethan house. A 50-foot pear tree. Some of us had been aware of the perry pear tree for a number of years. I knew that it was a superb specimen. Not only had we made perry with the pears and found it to be fantastic, we had taken steps towards identifying it. With the help of Liz Copas and East Malling we had ascertained that its DNA was unique. That it was a very rare tree in Somerset terms. Unique means unique. Nothing else like it. It was about 200 years old and in very good health for a perry pear and would have gone on for another 100 years at least.

Then, one Sunday morning, the chainsaws were heard,

hard at work. All the branches were lopped off cack-handedly till it looked like something from the Western Front. The reason for massacring the tree? Simply that every autumn its leaves were falling on the young lad's revamped polo field. I tried to get a Tree Preservation Order on what remained of the tree but the local tree officer from the council was less than helpful. In fact, he was not in the slightest bit interested. He laughed – not a good move. He said it would cost taxpayers' money . . . which, to my mind, shows a complete and utter ignorance of the problem. Other perry pear specialists further afield were horrified when they saw the result of the massacre. Mature perry pear trees in Herefordshire and Gloucestershire are respected in those counties. If we are to have good trees in the future they must be protected and conserved. Pears for heirs. Preferably heirs without chainsaws. And that was the only tree of its kind not just in Somerset but in the whole world. Rarer than white rhinos. Yet it was on our doorstep. Criminal . . .

Liz Copas and another local resident took some cuttings from what was left on the bonfire and we hope with the help of a local tree nursery in East Devon to be able to resuscitate the tree. But it will take fifty years to get a good crop. It was a salutary warning of the ignorance about certain trees in certain quarters. The future is conservation not desecration. Each county should perhaps form its own cider and perry pear tree association to register what is out there and make sure that the tree officer knows what you are doing. And, if necessary, slap a TPO on those trees. They are like gold dust. Tree officers are supposed to protect veteran trees, not turn a blind eye when they are either felled or massacred. Some of my friends who are proper tree surgeons would have refused to do such a job. The tree surgeon in question did a hurried, botched job so that he could get to a wedding on time. Unbelievable . . . the tree has since been registered and I hope it recovers from its ordeal. As for polo, that is another story, but I was assured that in the autumn there were still plenty of leaves on the field from other trees . . . Conservation is vital. In the next village there is another unique perry pear tree of similar age and that is doing fine. They are making perry from it. Sensible fellows.

Climate change

Climate change: the great unknown. Top fruit like apples and pears are fairly resilient and the seasons usually even out. Orchards often have 'on' and 'off' years. But orchards are now having a few problems. When I was last in Herefordshire I visited James Marsden at Gregg's Pitt Cider. We had a wide-ranging chat. James used to be in the upper echelons of Natural England.

James is an avid record keeper and gets out his cider and perry files: he is very proud of his scientific observations. It is a bit like being back in the early days of the Royal Society. James shows me a whole range of graphs, the visual manifestation of climate change. Impressive, hardcore evidence. There is even a great spreadsheet going back at least fifteen years.

'I wish I had kept it for the full twenty-five years. It shows variety, date pressed, original gravity, whether I added any sulphur dioxide and if so how much. The date of first racking, date of second racking, date of bottling. The gravity at bottling. It's all there. So that's my IP if you like.' James is meticulous. His cider and perry are excellent. As we talk James gives me a sample of his 40 percent abv pear spirit. Eau de vie distilled from his perry by Charles Martell down the road. Excellent, fresh and with great vigour. Almost crisp.

James has also noticed differences between old and young trees. 'Fruit quality is always better from the older trees without exception and I have now got the data to support that. With young trees the quality and depth are not there and the sugar levels will not be as high and that is important.'

Trees are cropping much earlier these days. which is both good and bad. Hard, long winters, which trees ideally need for dormancy, are now a rarity. James's graphs show at least three or four things: first, pressing dates are coming forward by as much as two or three weeks, the specific gravity, i.e. the sugar content, is going up because of the sunshine levels and acidity is in the middle range. This is good news. But because the pressing is early and the temperatures are higher, in September rather than late October, fruit, particularly perry pears, goes rotten much more quickly. As we have seen, they rot from the inside out. Use it or loose it. Hands-on, particularly during that first three to four weeks of the season otherwise you will potentially lose big volumes. You have to pick fruit almost every other day, which is real dedication. And when perry pears drop 50 feet you wear a helmet.

On my way back down to Somerset I also visited old friends, Norman and Ann Stanier, at Dragon Orchard in Putley. They have 15 acres of cider fruit and all the trees have just been pruned in three days by a gang of Eastern Europeans. Norman is very impressed. 'Six of them do what would take us three months. They start work at first light and work till dark, they do it brilliantly.'

Norman's family have been around for quite a while. His great-grandfather was overseer of the Putley Court estate in the 1850s when Squire Riley planted model orchards. 'Squire Riley went over to America and saw the big orchards in the Hudson Valley. His small, mixed farms soon became apple orchards. Fruit was sent direct to London from the local station three miles up the road by GWR and it would arrive early next morning ready for sale in Covent Garden. Those were the days.'

I ask Norman about global warming and he immediately mentions new orchard trials being conducted down at Reading University inside vast polytunnels where they can control temperature and humidity. Sounds very interesting. 'They are trialling different varieties in different conditions to see which ones will do well in the future. Lack of dormancy is a problem. Trees need to recover just like cider makers after a long, hard season. Also, if blossom comes early and is sporadic, germination is down. You need the bees. And if bees don't come out at the right time, pollination can be down by fifty per cent. Strong, cold winds from the north or east don't help either.' Climate change is a curate's egg.

What also impressed me was that Norman and Ann Stanier had pioneered an orchard-share scheme where people paid a certain amount per year and came to the farm from all over the country, four times a year, and took away cider, apple juice and fruit. There were talks and lectures, even poetry workshops and musical events. And a straw bale theatre, weddings, too. It was very successful. They called it Cropsharers. 'We ran that till 2016. Then we retired and went round the world for a year.'

'Orchard philosophy? On Norman's mother's side we have been in the village for nearly three hundred years. Makes us take the long-term view. In some ways purist. We like to learn from tradition but look forward. We make cider with Simon Day, who brought a wine maker's approach to cider. His father was involved with Three Choirs Vineyard and Simon made wine and cider in Jersey, at La Mare, and in New Zealand. He produces a very distinctive drink. So we have the best of both worlds.'

Anne won the *Country Living* Rural Business Woman of the Year Award a while ago. She was recognized nationally. 'We like to do things traditionally, middle path, our Herefordshire roots. Life is quieter now.' One key element of the future is making orchards more available to the public, just as they have at Hartpury with National Perry Pear Centre and at Brogdale in Kent.

Climate change is not all bad, but you have to think one or two steps ahead and be aware of annual fluctuations. As always, it comes down to timing. And with perry you have to be spot on.

The future

The future of cider is difficult to predict. Cider making, like any other business, is at the mercy of market forces. However rosy and romantic a cider farm may appear from the outside, with blossom and apples, there are always cycles of supply and demand, boom and bust. Throughout history there have been good and bad periods, but cider has a way of surviving because it brings people together. It is part and parcel of rural life. It was wages, it was life itself. As it happens, cider often does well in a recession. It is cheap and handy.

Cider history has its own roller coaster. Who could have predicted the Civil War in 1640s England? The punitive

Fig. 10.7 'Cider is Cider', Burrow Hill, 2020

cider tax of 1763? The emergence of factories in the nineteenth century or television advertising in the twentieth? Or the Internet? These 'advances', combined with new trends in cider drinking, like fruit ciders and pear cider, have had knock-on effects on traditional orchards. More recently the downstream effects of Covid and the absence of summer festivals in 2020 meant that thousands of tons of cider apples had to go into biodigesters to make methane. All gas and gaiters. But cider will recover. Maybe distilling is the way forward?

The secret, I suspect, is to make cider to the very highest standards with the best cider apples within your grasp and market it as such. You have to reach out and create those markets for yourself, not just to replace beer in the pubs and drinking dens or to imitate wine on the top table. Cider has to stand on its own two feet and compete on its own terms. It has everything going for it. A long, colourful, intrepid history, fantastic provenance and terroir, as well as elegance and aristocratic sparkle. It is infinitely flexible. Maybe it is time for cider to be taken for what it really is: a biologically diverse and sustainable drink with an incredibly long and interesting geographical and intellectual pedigree. Its history and connection to the land in this country is second to none. Cider is cider.

Even during the Covid crisis most artisan cider farms did quite well. During the long, hot, isolated summer of 2020, the country was very thirsty. Smaller cider makers sent regular shipments of top-notch bottled cider all over the country and into beleaguered cities. They did very well. Some even ran out. As always, distribution is the key. As well as the Internet.

Medium-size cider makers encouraged people onto their farms to walk around their orchards, to have a meal and a picnic. Families with children had time to absorb the magic of the orchards and make a deeper connection than by just driving into the farmyard and then driving out again. Some cider farms like Perry's in Somerset have farm shops and cafés. These were a lifeline for local people who could not go into supermarkets. Wood-fired pizza ovens sprouted up in many orchards. Sheppy's also have a restaurant, and farm shop. Their car park always seemed to be full. Rich's do well with their restaurant. Their pensioner lunches are very good value. I've had one. They have adapted to survive. Ross-on-Wye Cider even has its own pub, the Yew Tree at Peterstow. Their music evenings are wonderful, laden with hillbilly music, blues and country and western. The secret is to make the cider farm the hub of the local community in whatever way you can. Apple days and wassail are great. One farm at Brent Knoll has a wassail with belly dancers from Cheltenham, Morris men and mummers from Langport, hog roast and mulled cider. John Harris has raised money for cancer charities and the local primary school. Cider also fits very well into the Slow Food movement and makes its own connections on a very deep level. Cider is ancient and yet modern. It draws people together.

Raising the profile of cider both locally and in cities never goes amiss. Tastings are vital, whether it is at the farm, on stalls at farmers markets, at trade shows or in tutored tastings at food and drink festivals. The most important thing is to engage with the public at almost every level. And cider is incredibly flexible, going from farmhouse cider right through the bottle range to sparkling cider and spirits. All from the same orchard. Remarkable when you think about it.

For the general public the secret is to go out there with an old-style map and find these cider farms, driving down those long, winding lanes in the back of beyond. Make an adventure out of it. Sometimes there is just a white arrow in a hedge on a small lopsided board with the word 'Cider' written on it. That's enough. Just go for it . . .

Everything hinges on orchards. Cider makers have to look thirty or even fifty years ahead. No easy matter. The investment in planting an orchard is high both in work, time, money and commitment. You need acres, good soil, good climate, terroir, good stories and good publicity. Land is never cheap. But once you have an orchard it will keep you company all your life. You can even sling a hammock between two trees. Or run it as a camping site. You also have to know what sort of cider you want to drink before you plant the trees. Cider research is vital. And that means talking to cider makers, sampling, tasting, taking notes if need be, honing your palate, building up a library of bottles and books.

What the future holds for large companies I am not sure. The rapid demand in the public's taste for fruit ciders means that some recently planted large-scale bittersweet orchards have been grubbed out. Also some new perry pear orchards have been grubbed out as the demand for perry pears has dwindled, which is a terrible shame. Pear cider, the oxymoron, came into this country with Kopparberg from Sweden in 2003. If you want to conserve old pear trees, some of which are 200–300 years old, seek out the perry makers, taste their perry. Sparkling bottle-fermented perry is fantastic for weddings. A right pear.

For a glimpse of the future you can always visit The Newt in Somerset at Hadspen House, between Castle Cary and Bruton. Nothing can prepare you for the elegant opulence of the surroundings, the intricately landscaped garden that Batty Langley would have been proud of, as well as modern, state-of-the-art, all-singing, all-dancing cider works which looks more like something out of the engine room of a spacecraft. Not a cobweb in sight. Only pipes and stainless-steel vats, large and small, as well as a bit of oak, with all sorts of wizardry to control the temperature. Slow, cool fermentations are the order of the day. No accelerated, high-speed ferment or codswallop cider here. Everything seems to be computerized and under control from the command centre high above the shop floor. New techniques have come from the wine industry. The Newt is linked to Babylonstoren, a state-of-the-art wine farm in the Western Cape. Wine and cider linked at the hip.

Hadspen House is the old demense of the Hobhouse family. The Georgian mansion is now a hotel. There is a small village with shops and restaurants, bakeries, butcheries, delicatessens and market gardens. More like a modern villa for a consul or senator. Somerset has not seen anything like this since Georgian Bath was built and cider, not wine, is centre stage. When in Rome do as the Romans do. When in Somerset make cider. And their cider is made from local cider apples: Kingston Black, Dabinett, Yarlington Mill and Somerset Redstreak. Distilling is now on the cards. Even Julian Temperley is impressed and will help make Newt cider brandy – a joint venture.

To have such a large park and gardens open to the public and to make cider centre stage is real statement of intent. The owners, cider makers and cellar masters are obviously in for the long term. The Newt is very happily swimming in cider. They have just launched their own traditional method sparkling cyder called 'The Winston'. The future is bright. At The Newt cider makes it own waves.

Conclusion

Across the country cider is in very good shape. New cider makers have popped up not just in every county but within cities as well. This widens the base outside the traditional West Country cider counties. But cider made from dessert apples in the east of England is a very different beast from the bittersweet apples of cider country. Real artisan perry needs a bit more help. Plant perry pears and perry orchards for 2050. Pears for heirs. The only real problem on the horizon is climate change and the way in which trees and fruits respond, as well as extra pests and diseases. Think strategically. Think ahead of the curve if you can. People are already planting cider apples in Scotland, Norway and Sweden. Even Iceland. The trick is to get the right variety that is disease-resistant and one that feels at home in the local climate and survives hot summers and wet winters.

As for cider history, it has come full circle. The site of Lord Scudamore's house at Holme Lacy, near Hereford, is now a Warner Hotel with a swimming pool. John Beale's name is still on the church wall in Yeovil. Ralph Austen's house in Oxford, where he made his cider – 30 Queen Street – is now a Costa Coffee shop, which might have amused him. In Petersfield John Worlidge's home – Worcester House, 2 & 4 Dragon Street, is now a holistic wellness centre, which is appropriate. As for Hugh Stafford, Pynes House is a wedding venue. The large orchard just hanging on by its fingertips. I recently spent a wonderful morning there with Polly Hilton and Liz Copas trying to identify cider apples. Half the trees were falling over. Desperately in need of attention. Find & Foster.

We all need experts and expertise from time to time. Drink full juice cider, that is the only real way forward. That way you sustain English cider orchards. Changing labelling rules is perhaps a vital step forward so the cider drinker knows far more about their cider: apple juice content, sweeteners, colouring, that sort of thing. Knowing how much sugar is in there is crucial. Then seek out artisan makers and understand their cider apples, their tannins, their passion. Go to source, ask questions and get answers from the horse's mouth. Cider has a fine future, a strong, diverse history and a dedicated band of cider makers who spend their lives trying to make the best drink possible.

As for Andrew Lea, the cider expert keen on tannins, his trip to Kazakhstan had left a very deep impression. As he said, 'It really hit me even more when I came home in September and I was gathering up apples from under one of my trees back in my orchard in Oxfordshire. I said to the trees: "Do you know I have just come back from meeting your cousins?"'

Fig. 10.8 'Harvest cider jug', made by Harry Juniper, Bideford, Devon, 2001

EPILOGUE

The world of cider apples is always in flux, always unfolding, evolving, moving on. It is often what you don't know that is most interesting. Discoveries can still be made. You have to keep your eyes open.

A year or two ago I was on a long 13-mile ramble in West Dorset. I visited a remote village and followed a footpath up a steep hill but it veered off into a thicket of brambles and there was no obvious way forward, so I climbed up the last bit of the slope beside the hedge through a large patch of nettles. Not pleasant, but all of a sudden I was assailed by a very familiar smell: the rich, aromatic perfume of mature cider apples. The ground smelled like a cider farm. For a few seconds I could not understand it but then I looked down. In among the nettles there was a carpet of bright red, crimson, almost purple apples. No idea what they were. I picked one up and bit into it. My mouth dried up. Massive hit of tannin. One bite, two bites, that was all, almost spat it out, but

I knew it was good. It was a cider apple for sure. But which one?

Cider makers always bite into apples and as you sink your teeth in you assess the texture. You taste the juice as it hits your taste buds. You register it on several different places in your mouth at several different levels. There is also the noise as the apple splits apart. Some apples are crisp, others spongy, a bit like cardboard. Astringent. Your tongue dries up. This is the mouthfeel that Andrew Lea talked about. An essential experience for cider makers. Almost addictive. These are the tannins you are after. They are the unsung heroes, the backbone of good cider. Always balancing flavours against your past encounters with apples. The smell of apples is a good indicator and I was not disappointed. You learn on the hoof.

I followed the red carpet and went up to the top hedge and a lone apple tree. The nettles were dying so the apples became more obvious. They were almost shiny from recent rain and still in very good condition. No sign of disease. Badgers and birds had left them well alone. I liked the taste very much. It was a bit like one of those early Celtic legends but there was no music and no young ladies either...

Luckily I had an old shopping bag stuffed into my jacket pocket so I filled the bag with apples and lugged them back home. When I got back I showed the apples to Liz Copas to find out what they were. They looked a bit like Kingston Black but I wasn't sure - and, anyway, what was it doing in a hedge? Even Liz didn't know what it was. She thought it might be Ten Commandments crossed with a crab. Or was it an unknown variety? She was perplexed. It tasted wonderful. If it was a 'new' variety that would be quite something. We went back and collected more apples. We pressed the apples and put the juice into demijohns. The juice was extraordinary. The apple was white inside with only the red on the skin. But the juice was a deep red, between strawberry and plum. It smelled very fruity and tasted superb. This was real scrumping. In the 1840s young lads were fined heavily for such crimes, and repeat offenders deported to Australia.

Nick Poole and Margaret at the West Milton cider club very kindly processed the apples for us and we waited to see how the cider progressed. Three or four bright red demijohns. Liz sent some of the juice off to Thatchers to be analysed and it came back with all the right indicators. Good acidity, good sugars, good tannins, good colour. Well-balanced. Just what you want for a single-variety cider. To get that depth of colour naturally was very unusual. We then let the wild yeasts work their magic and waited three months to see if the cider was any good. A slow process. In the meantime, Liz sent samples of leaf off to get the DNA tested at East Malling Research Station in Kent.

When the results came back it was deemed 'unique'. A brilliant result. The cider was very dry with a high level of tannins. Just what I liked and still bright red. Very unusual indeed. It reminded me of a wine I had tasted called Madiran from southwest France made with the Tannat grape. High tannin and pretty challenging. This was the cider equivalent or at least halfway there. It would be interesting to test the cider alongside traditional varieties like Harry Master's Jersey or Tremlett's Bitter.

One of the great things about discovering a new cider apple is that you get to name it. The name Rough 'n Reddy jumped out at me. The apple was very red and very rough. Some might say that the cider being high tannin was only good for blending but I liked it on its own. It got the thumbs-up when we tasted it at West Milton cider club. Later on Liz Copas registered the apple and it is now on the Fruit Forum National Database.

In fact, it produced a superb cider, a deep, dark red which kept its colour and tasted almost plummy. Wonderful deep, earthy aromas. Very rough and very reddy. But difficult to propagate. A cider maker's cider. Never mainstream but a valuable addition to the Dorset cider library. My own small foray into wild cider. A chance discovery. Almost as good as a trip to Kazakhstan.

Fig. 11.1 'Cider Country', Clifford Lugg and his son Barry drinking cider at Tregarne Farm, St Keverne, Lizard, Cornwall, 1978

ACKNOWLEDGEMENTS

Writing *Cider Country* has been yet another wonderful voyage into the magical world of cider and cider apples where orchards and cider farms are constant companions. Research is always interesting, and I am very grateful to those who gave freely of their time and expertise, in particular Andrew Lea and Liz Copas both of whom worked at Long Ashton. Then Elizabeth Pimblett, the Director of the Museum of Cider in Hereford, and Sally Mansell the archivist. Thanks also to all the librarians and archivists at the Royal Society who, back in 2008, let me see the original material from John Evelyn's *Sylva* and *Pomona* from the 1660s.

Thanks to all the cider makers: Julian and Matilda Temperley, Burrow Hill and Somerset Cider Brandy Company; Kingsbury Episcopi; Alex Hill of Vigo and Bollhayes Cider, Clayhidon' Tom Oliver of Ocle Pychard; James Marsden of Gregg's Pitt; Much Marcle, Mike and Albert Johnson of Ross on Wye Cider, Peterstow; Paul

Stephens of Newton Court Cider, Leominster; Norman and Ann Stanier, Dragon Orchard, Putley; Nick Poole of West Milton Cider, Powerstock; Joe Heley of Secret Orchard, Nettlecombe; Polly and Matt Hilton of Find and Foster, Huxham Barns; Vernon Shutler of Countryman Cider Felldownhead, Milton Abbot; George Perry of Perry's Cider, Dowlish Wake; Louisa Sheppy of Sheppy's Cider, Bradford on Tone. And a very wide range of other cidery contacts: John Cluett, ex-truck driver and squeezebox player from Shaftesbury. Sue Clifford and Angela King, *old* Common Ground, also from Shaftesbury. Anthony Gibson of Langport. Historian and classicist Bijan Omrani of Shute and Eddie Smith at Westminster School. Richard Watkins of Barrington for supplying me with old cider books. Michael Pidd the Director at the Digital Humanities Institute, University of Sheffield for permission to use the Samuel Hartlib material. Archaeologists Peter Addyman in York and Tom Greeves of Tavistock; Sigrun Appleby/Hólmsteinsdóttir from Reykjavík and Sæmi - *Sæmundur Guðmundsson* from Hella in Iceland; Oake Parish Council for details about Heathfield Rectory and early days of Taunton Cider. Julia Coutanche of Jersey Heritage Trust in St Helier for all her help over the years investigating Jersey Cider. Prof Jo Story of University of Leicester for referencing details of Charlemagne's edicts in *Capitulare de Villis.*

I am also grateful to The Museum of Cornish Life, Helston to be able to use the photographs of cider making in the Lizard. And to Cornish pomologists James Evans and Mary Martin, Chris Groves from National Trust, Robert Dunning, Somerset County Historian, Tom Jaine, Allaleigh and

Prospect Books and Sara Hudston for the tip-off about Nart Sagas. Also, my excellent agent Jessica Woollard of David Higham who definitely likes good cider and my eagle-eyed editor Grace Pengelly at HarperCollins who keeps me up to scratch. And finally, Carla my wife for helping sample so many excellent ciders and perries along the way.

BIBLIOGRAPHY

Atherton, Ian, *Ambition and Failure in Stuart England, The career of John First Viscount Scudamore*, Manchester University Press, 1999

Austen, Ralph, *A Treatise of Fruit Trees*, with *A Spiritual use of an Orchard*. Henry Hall, Oxford, 1657

Baring Gould, Sabine, *A Book of Devon*, Methuen, 1899

Barty-King, Hugh, *A tradition of English Wine Making*, Oxford Illustrated Press, 1977

Beale, John *Herefordshire Orchards, a Pattern for all England, written in an Epistolary Address to Samuel Hartlib, Esq.* London. 1656; reprinted in Dublin 1724.

Bellamy, D, *Nature Delineated Vol 1 and 2* London 1739

Bellamy, Liz, *The Language of Fruit*, University of Pennsylvania Press, 2019

Brown, Pete, *The Apple Orchard*, Particular Books, 2016

Browning, Frank, *Apples*, North Point Press, New York 1998

Cell, Gillian T, *Newfoundland Discovered, English*

Attempts at Colonisation 1610-1630, The Hakluyt Society, London, 1982
Chapman, Jim The *Cider Industry and the Glass Bottle,* Gloucestershire Society for Industrial Archaeology Journal for 2012 pages 36-40
CIVC - Comité Interprofessionel du Vin de Champagne, *Champagne,* Epernay, 2007
Common Ground, *Apple Games & Customs*, Common Ground, 2005
Common Ground *The Apple Source Book* Common Ground, 1991
Common Ground, *Orchards, A Guide to Local Conservation*, Common Ground, 1989
Common Ground, *The Common Ground Book of Orchards*, 2000
Colarusso, John, *Nart Sagas: Ancient Myths and Legends of the Circassians and Abkhazians* Princeton University, 2016
Copas, Liz, *A Somerset Pomona*, Dovecote Press Wimborne, 2001
Copas, Liz, *Cider Apples The New Pomona,* 2013
Crowden, James, *Cider: The Forgotten Miracle*, Cyder Press 2, Somerton, 1999
Crowden, James, *Ciderland,* Birlinn, Edinburgh, 2008
Crowden, James *Dorset Man*, Agre Books Dorset, 2005
Crowden James, *Dorset Women*, Agre Books, Dorset, 2006
Crowe, William, *Lewesdon Hill*, Clarendon Press Oxford, 1788, reprinted Flagon Press, 2007
Cunliffe, Barry, *The Scythians*, OUP, 2019

Customs and Excise, HM, *Cider and Wine Production Notice 162*, 2002

Defoe, Daniel, *A Tour Through the Whole Island of Great Britain,* 1724-1727

Digby, Sir Kenelm, *The Closet of the eminently learned Sir Kenelm Digby Opened,* London 1669 reprinted Prospect Books, 1997

Di Palma, Vittoria, *Drinking Cider in Paradise: science, improvement, and the politics of fruit trees,* Chapter 10 of *A Pleasing Sinne: Drink and Conviviality in Seventeenth-century England* edited by Adam Smyth Cambridge, 2004

Dovaz, Michel, *L'encyclopédie des vins de Champagne,* Julliard, 1983

Elville, EM, *English Table Glass*, Country Life, 1951

Evelyn, John, *Sylva* with *Pomona*, and *Aphorisms concerning CIDER* Royal Society, 1664

'E.T' *The Art & Mystery of Vintners and Wine Coopers,* 1675

Fiennes, Celia, *The Journeys of Celia Fiennes,* Cresset Press, 1947

Finberg, HPR. *Tavistock Abbey*, David & Charles, 1969

Foot, Mark, *Cider's Story Rough and Smooth,* Nailsea, 1999

Fagan, Brian, *Fish on Friday,* Basic Book, 2006

French, Roger K, *The History and Virtues of Cyder,* Robert Hale, 1982

Gee, Michael, *The Devon Orchards Book*, CPRE/ Halsgrove, 2018

Gerard, Thomas of Trent, *The Particular description of*

the County of Somerset, 1633, reprinted Somerset Record Society, Edited EH Bates 1900 & Castle Cary Press, 1973

Godfrey, Eleanor S, *The Development of English Glassmaking 1560-1640,* Clarendon Press Oxford, 1975

Hartschorne, Albert, *Old English Glasses*, Edward Arnold, 1897

Heywood, Audrey *Effect of Immersion on Urinary Lead Excretion.* British Journal of Industrial Medicine, 1986;43:713-715.

Heywood, Audrey, *A Trial of the Bath Waters: The Treatment of Lead Poisoning,* Medical History, Supplement No. 10, 1990, 82-101

Hogg, Robert, & Bull, Henry Graves, *Herefordshire Pomona.* Two Vols Woolhope Naturalists Field Club, 1878-1885

Jeffreys, Henry, *Empire of Booze,* Unbound, 2016

Juniper, Barrie E, and Mabberley, David J, *The Story of the Apple,* Timber Press, 2006, 2nd Edition Kew Publishing, 2019

King, Angela, & Clifford, Susan, *Holding Your Ground*, Wildwood House, 1988

Knight, Thomas Andrew, *A Treatise on the Culture of the Apple and Pear and on the manufacture of Cider & Perry.* Third Edition H. Proctor Ludlow, 1808

Langley, Batty, *A Sure Method of Improving Estates,* 1728

Lawson, William, *A New Orchard and Garden*, 1618, reprinted Prospect Books, 2003.

Le Paulmier, Julien, *De vino et Pomaceo- Le Premier Traité du Sidre,* 1588

Lea, Andrew, *Craft Cider making*, Good Life Press, 2008

Legg, Philippa, and Binding, Hilary, *Somerset Cider,* Somerset Books, 1988

Loyn, HR and Percival, John, *The Reign of Charlemagne. Documents on Carolingian Government and Administration,* Edward Arnold (London, 1975), pp. 64–72.

Luckwill, LC & Pollard A, *Perry Pears,* University of Bristol, 1963

Mattson, Henrik, *Calvados,* Flavourcider.com, 2004

Meacham, Sarah Hand, *Every Home a Distillery*, John Hopkins University Press, 2009

Merrett, Christopher, *The Art of Glass - Antonio Neri* 1611 - translated from Italian and printed by A.W. for O. Pulleyn in London England, 1662

Mabey, Richard with Susan Clifford & Angela King for Common Ground, *Second Nature*, Jonathan Cape, 1984

Maclaran, Pauline & Stevens, Lorna *Magners Man: Irish Cider, Representations of Masculinity and the Burning Celtic Soul* Irish Marketing Review Volume 20, No 2, 2009

Morgan, Joan, *The Book of Apples*, Ebury Press, 1993

Morgan, Joan, *The Book of Pears*, Ebury Press, 2015

Mortimer, John, *The Whole Art of Husbandry*, London, 1707

Nabhan, Gary Paul, *Where our Food Comes from -*

Retracing Nikolay Vavilov's Quest to End Famine, Island Press/Shearwater Books, 2009

Nye, Robert, *Beowulf a New Telling,* first published as *Bee Hunter: Adventures of Beowulf,* Faber and Faber, 1968

Parkinson, Anna, *Nature's Alchemist, John Parkinson herbalist to Charles 1st,* Frances Lincoln Limited, 2007

Parkinson, John, *Paradisi in Sole Paradisus Terrestis*, London, 1629 reprinted Methuen, 1904

Penn, Herbert, *Glassmaking in Gloucestershire,* Journal of the Gloucestershire Society for Industrial Archaeology, 1983, pp. 3-16, ill.

Philips, John, *CYDER - A Poem*, Blackfriars, 1708

Philips, John, *CYDER - A Poem*, with extensive notes provincial historical and classical by Charles Dunster, London, 1791

Potter, Jennifer, *Strange Blooms, The Curious Lives and Adventures of the John Tradescants*, Atlantic Books, 2006

Potter, Jennifer, *The Jamestown Brides*, Atlantic Books, 2018

Powell, Harry J, *Glassmaking in England*, Cambridge University Press, 1923

Raymond, Walter, *Good Souls of Cider-Land*, Grant Richards, 1901

Roberts, Alice, *Tamed, Ten species that Changes our World*, Windmill Books, 2017

Reiss, Marcia, *Apple*, Reaktion Books, 2015

Roach FA, *Cultivated Fruits of Britain*, Basil Blackwell, 1985

Simon, André, *Drink,* Burke Publishing, 1948

Spengler III, Robert N, *Fruit From the Sands, The Silk Road- Origins of the Food We Eat,* University of California Press, 2019

Spiers, Virginia, *Burcombes, Queenies and Collogetts*, West Brendon, 1996

Stafford, Hugh, *A Treatise on CYDER-MAKING,* London, 1753

Stevenson, Tom, *World Encyclopaedia of Champagne Sparkling Wine,* Absolute Press, 1998

Trewin, Carol, *The Devon Food Book*, Flagon Press, 2010

Wallace, T & March RW, Ed, *Science and Fruit*, LARS University of Bristol, 1953

Watts, David C, *The History of Glassmaking in London*, Watts Publishing, 2009

Whiteway, EVM, *Whiteway's Cider*, David & Charles, 1990

Wilkinson, LP, *Bulmers of Hereford*, David & Charles, 1987

Williams, Ray R, *Cider and Juice Apples*, Edited, University of Bristol, 1996

Wilson, C Anne, *Water of Life*, Prospect Books, 2006

Worlidge, John, *Vinetum Britannicum or a Treatise of CIDER*, Thomas Dring, London 2nd edition, 1678.

LIST OF IMAGES

Introduction

1.1 'Cidermaking in the 17th century was a gentlemanly art', (1678) from *Vinetum Britannicum* Or, *A Treatise of Cider*, 2nd edition by John Worlidge, London, Thomas Dring, (1678).

Chapter Two

2.1 'Route of Alexander the Great's apple expedition to the Middle East and part of Central Asia' (*c.* 334–323 BC) *Geography of the Ancients*, 5th Edition printed for Christopher Browne, London, (1725).

2.2 'Autumnus, Pomona's sister' (4th century AD) Roman Mosaic in a bathhouse, Villa Las Tiendas, Spain Alamy Stock Image ID P79B6H.

Chapter Three

3.1 'Freia: a combination of Freyja. Goddess of love, beauty and guardian of the golden apples and the goddess Iðunn', Illustration by Arthur Rackham for Richard Wagner's *Das Rheingold* (1910) reprinted (1939).

Chapter Four

4.1 'Ancient olive press, with windlass', Arboretum Tresteno, Croatia' (17th Century) © Miomir Magdevski, Serbia. Image reproduced under the Creative Commons Attribution-Share Alike 4.0 International license.

4.2 'Ancient & Modern' Clifford Lugg and his son Barry making cider at Tregarne Farm, St Keverne, Lizard, Cornwall. HESFM:2016.13284.1 © Museum of Cornish Life, Helston (1978).

Chapter Five

5.1 'Perfect forme of a fruit tree', (1618) William Lawson, *A New Orchard and Garden* (1618) & Prospect Books, (2003).

5.2 'Yorkshire Eden. Orchard & kitchen garden with still houses and beehives in all four corners', (1618). William Lawson, *A New Orchard and Garden* (1618) & Prospect Books (2003).

5.3 'Orchard Portman and its extensive orchards', Taunton, Somerset (1707) From an engraving by Pieter Van der Aa after a painting by Leonard Knyff (1650–1722) now held in the Royal Collection Trust (*c.*1705).

Chapter Six

6.1 'Barent Langenes map of Terra Nova', Newfoundland (1602) Digital Archives Initiative, courtesy of the Centre for Newfoundland Studies, Memorial University of Newfoundland, St. John's, NL Canada.

Chapter Seven

7.1 'Sir Kenelm Digby, diplomat, courtier, naval administrator, alchemist, pirate, glassmaker, cidermaker, food and drink historian' (1654) Engraving after Sir Anthony van Dyck.

7.2 'Lord John Scudamore, MP, diplomat & cidermaker' (1601-1671) Image taken from *Herefordshire Pomona* (1878).

7.3 'Lord Scudamore's flute also known as the Chesterfield flute, a soda cristallo drinking glass, 14 inches high' (*c.*1640)' © Museum of London 34.139/1.

7.4 'Walled garden with fruit trees', from *A Treatise of Fruit Trees* by Ralph Austen (1657) Printed by for Thomas Robinson, Oxford (1657).

7.5 'The 'Busby' cider bottle', recently excavated at Westminster School', London (*c.*1665). Source article: 'Onion-shaped bottle from the Great Kitchen, Westminster Abbey site', Eddie Smith (2017).

Chapter Eight

8.1 'No Excise Georgian cider glass' (1763) Sold for £5625 at Bonhams London, Lot 67, 13 Nov 2013. Image courtesy of Bonhams.

8.2 'The Ingenio', A robust scratter for crushing cider apples, no need of a horse', from *Vinetum Britannicum Or, Treatise on Cider* by John Worlidge Gent, printed for Thomas Dring, London (1678).

8.3 'Primitive but effective distilling apparatus' from *The Art of Distillation* by John French (1651).

8.4 'John Philips, The Cider Poet', Unknown Artist (*c.*1700) © National Portrait Gallery, NPG1763.

8.5 'No Cyder Act Teapot', Yorkshire Creamware with crab-stock spout and reverse inscription 'Apples at Liberty' (*c.*1766) Sold by Bonhams, Oxford, for £5,000 Lot 27, 16 Jan 2013. Image courtesy of Bonhams.
8.6 'The Cider Monument' Burton Pynsent, Curry Rivel, Somerset. Designed by Capability Brown, Portland Stone, 140 ft high (1767) Photo © James Crowden.
8.7 'Detail of 'No Excise On Cyder', eight day long case clock made by Joel Spiller of Wellington (*c.*1763) Photo courtesy of Bonhams and Julian Temperley, Burrow Hill Cider Farm, Kingsbury Episcopi, Somerset.
8.8 'Thomas Andrew Knight, horticulturalist, botanist and cidermaker' (1759-1838) from *Herefordshire Pomona*, (1878).

Chapter Nine

9.1 'Robert Neville-Grenville, Squire of Butleigh in Somerset, looking very dapper' (1906) Courtesy of South West Heritage Trust SHC D/P/butl/23/4 and Butleigh Parish Collection.
9.2 'Nigel Stewart of Bridge Farm Cider explaining the finer points of cider making', Covent Garden, Apple Day (1990) © Common Ground.

Chapter Ten

10.1 'The Big Squeeze - Building the big cheese', Apple Day, Cotehele House, Cornwall (2019) Courtesy of The National Trust. Photo © James Crowden.
10.2 'Cidermaking at Westcott Farm', near Callington,

Cornwall (1981). Oil painting by Mary Martin. Also features in *Burcombes, Queenies and Collogetts,* West Brendon (1996) Courtesy of the National Trust who now own the painting. On display at Cotehele House. NT 348090.

10.3 'Bottle fermented sparkling cider maturing in a *pupitre* or 'A' frame', Burrow Hill Cider Farm, Somerset (2021) Photo © James Crowden.

10.4 'Julian Temperley in the distillery', Somerset Cider Brandy Company (2020) Photo © Matilda Temperley

10.5 'Julian and Di Temperley in their bonded warehouse. Cider brandy well matured - oak matters' (2020) Photo © Matilda Temperley.

10.6 'Polly Hilton in old orchards', Upton Pyne, Devon (2020) Photo © James Crowden.

10.7 'Cider is Cider', Burrow Hill (2020) Photo © James Crowden.

10.8 'Harvest cider jug', made by Harry Juniper, Bideford, Devon (2001) Photo © James Crowden.

EPILOGUE

11.1 'Cider Country', Clifford Lugg and his son Barry drinking cider at Tregarne Farm, St Keverne, Lizard, Cornwall. HESFM:2016.13284.6 © Museum of Cornish Life, Helston (1978).

INDEX

Page references in *italics* indicate images.

Achilles 40
Act of Supremacy (1534) 114
Adams, John 160
Adam's Pearmain 103
Addyman, Peter 73
Adelard 111–12
Admiralty 148, 149, 219
'Adventure of Conle, The' 78
'Afallenau' (poem) 79
Afghanistan 8, 17, 18–19, 45
Aghra, Ailill Ochair 76–7
Aillin 78
alchemy 100, 112, 115, 127, 175, 176, 177, 178, 204, 242
Alcinous, King 42
Alcmene 40
Alexander the Great 14, 31, 44–5, 46, 47–50, *48*, 240
Alexander, Pope 104
Allen, Thomas 175, 177–8
Almaty, Kazakhstan 15, 16, 17, 19, 20, 21, 24, 25, 26, 31
Ameline, Michel 351
America 154, 155–63, 164, 165, 166, 265, 269–70, 275, 276–7, 309, 334, 337, 346, 365, 392–3
    American Revolution (1765–83) 270, 276
Anaxarchus 49–50
Androclyes 240
Anne, Queen of Great Britain 247, 248, 254
Anson, George 144
antiquaries 116, 117, 120–3, 127, 133
Aphrodite 40, 41, 43, 44, 46
Apian 54
Apicius 56
Appius, Claudius 54
Appleby, Sigrun 67, 415
Apple computer logo 64
Apple Days 336–8, *340*, 342–5, 349, 355, 360, 362, *363*, 403
apple detectives 359–66
apple dining clubs 5, 53–6
apple forests 8–9, 11–30, 31, 85
apple identification 68, 302, 355–9
apple throwing 44–50
apple trains 310–12
apple wine (*æppelwīn*) 88
Arabella 159
archaeology, apple 12, 72–3
Archimedes 50–2, 53, 91, 210
Archimedes screw 51–2
Archytas of Tarentum 90
Argentocoxu 81

Aristophanes: *Clouds* 46–7
Aristotle 14, 44, 45, 45*n*, 124
Ark of Avalon 154
Armada, Spanish 138, 139, 152, 153, 386
Armeia 54
Armenia 13, 36
arstibara ('Apple Bearers') 49
Arthur, King 84–5, 87, 88
artisanal cider 7, 37, 71, 89, 169, 239, 252, 279, 300, 326, 328, 330, 333, 334, 336, 370, 379, 390–4, 402–3, 406, 407
Arundel, Thomas 181–2
Aspalls Cider 316
Atalanta 40, 43–4
Athena 40–1
Athenaeus: *Deipnosophistae* 38
Athenian wedding feasts 46
Aubrey, John 135
Austen, Ralph 190–1, 195, 198, 202, 204, 207, 209, 217, 232, 362, 407; *A Treatise on Fruit Trees* 191–5, *191*, 229
Autumnus 59, *60*
Avalloc 84
Avalon 84–6, 121
Axmouth 104
Aÿ 220

Backamore 257
Bacon, Francis: *New Atlantis* 141
Bacon, Roger 112, 242
Baker, Sir George 281–3
Baltonsborough 364, 373
Banks, Joseph 290
Banwell, Andy 381
Barbary of Rever Jobles 107
Bardsey apple *Afal Enlli* 79–80
barilla 169–70
Barker, B. T. P. 305–6, 307, 335
Bartestree 379
Bateson, William 18
Bath & West Show 295, 302, 303, 316
cider competition 350, 385–6, 388
Battle Abbey, Sussex 103–4, 286
Beale, John 188, 196–9, 201–11, 212, 215, 220, 228, 229, 232, 297, 407; *Aphorisms on Cider* 209–10, 211
Beaulieu Abbey 105
Bedern 73
Benedict of Nursia 100
Benedictine monks 99, 100, 104, 220
Bennett, Richard 158
Beowulf 63, 120
Berkeley Hundred Colony 157
Berkeley, Richard 156–7
Biran, Matthew de 150
Black Book of Camarthen, The 79
Blackdowns 381, 386, 393
Blackstone, William 158–9
Black Tom 132
Blane, Sir Gilbert 148; *Observations on the Diseases of Seamen* 148
Bloody Butcher 361
Boethius: *The Consolation of Philosophy* 108
Bollhayes Cider 393
Bonal, François 224–5
Boorde, Andrew 116–17
Boston 158–9
Boston Tea Party 276
bottling cider 164–230
Bowe, Sir Jerome 166
Bradley, Natasha 381
Bran mac Febail 74–5
Bray, William 160
Breadfruit 361
Brereton, William 196
Bridge Farm Cider 340
Brittany 78–9, 83, 88, 94, 95, 311
Broadstreet Glass Works 169
Brogdale Horticultural Trust 251
Brogdale orchard 113, 229–30, 343, 366
broquelet 224
Brown, Capability 273, 274, 275, 277
Browning, Frank 6; *Apples* 22
Brunschwig, Hieronymus: *Liber de arte distillandi de simplicibus* 243
Buckingham, George Villiers, 1st Duke of 180, 181
Buckingham, 3rd Duke of 112
Buckland Priory 139
Bull, Edith 236
Bull, Henry Graves 236
Bulmer, Bertram 231, 234, 236, 298, 331, 372, 379
Bulmer, Revd Charles 297
Bulmer, Fred 234–5, 298
Bulmer, Gillian 236, 379, 395
Bulmer, HP 235, 297–8

Bulmers 52, 298, 300, 307, 308, 311, 312, 313, 314, 315, 316, 318, 328, 330, 331, 369
buoyancy 50–1
Burleigh, Sophie 380
Burrow Hill Cider Bus 86
Burrow Hill Cider Farm, Somerset 5, 8, 86, 277, 343, 372–8, *375*, *377*, *378*, 380, *401*
Busby, Dr Richard 199–200, *200*
Bute, Lord 260–1, 262, 263

Cabot, John 150
Cake, Penny 380
Calvados 241–2, 243, 343, 372–8, *375*, *377*, *378*
Camden, William 85–6; *Britannia* 120–1, 285–6
Cann Orchard 99
Carse of Gowrie 81–2
Çatalhöyük 12
Catlow, Revd William Edgar 296
Cecil, Robert 127
Celtic beliefs 57, 61, 62, 70, 71–2, 74–7, 80, 81, 82, 83, 92, 94, 97, 410
champagne 173, 179, 183, 188, 193–4, 201, 204, 219–30, 231, 235, 298, 367, 371
Champagne Bureau UK 227–8
Chapman, Jim 365
Charlemagne: *Capitulare de Villis* 97–8, 99, 111, 415
Charles I, King of England 128, 129, 175, 176, 180, 181, 182, 183, 184–5, 205
Charles II, King of England 174, 260, 286, 287
Charles the Simple 99
Charleton, Dr Rice 282
Charleton, Walter: *On the Adulterations of Wine* 208
Chaucer, Geoffrey: *The Canterbury Tales* 106
cheddar 121, 251, 256, 306, 373, 385
China 12, 16, 20, 27
Christianity 78, 94, 95, 97, 101, 114
cider banquet 379
cider bars 220, 369, 391
'Cider Bible' 106
cider brandy 112, 115, 130, 238, 241–2, 243, 343, 372–8, *375*, *377*, *378*, 380, 406
cider cellar 201, 231, 249
cider club 6, 209, 336, 347–9, 352, 354, 355, 411, 412
cider farm 1–9, 11, 34–5, 37, 47, 86, 134, 277, 296, 312, 315, 330, 350, 351, 357, 361, 372–8, *375*, *377*, *378*, 379, 392, 394, 401, 402, 403, 404, 409
cider festivals 82, 347, 349, 352–5
cider houses 34, 149, 155, 305, 306, 347–8, 374, 390
'Cidermaking at Westcott Farm', near Callington, Cornwall' *364*
Cider Monument 269, 273–7, *274*, 373
Cider Museum Trust 234
cider presses 3, 27, 50, 100, 102, 103, 104, 108, 115, 139, 155, 218, 225, 232, 309, 312, 316, 343, 347, 355, 379, 380, 381
cider riots (1763) 5, 231, 238, 262, 263–9, 276
cider vinegar 32–3, 134
Cider Women 379
cidre bouché 193, 351
Cirencester 60
Cirencester Agricultural College 383, 384
Cistercian monks 71
Citois, François 281
Claridge, John 359
Claudius, Emperor 92
Cleitus the Black 47–9, 50
Clifford, Sue 337, 338, 341–2, 343
climate change 397–400
Cluett, John 310–12
Cluniac Priory 115
Coates of Nailsea 296
Cockagee 257, 258, 291
Cockpit Improved 73
Cockrem, Mary Grace Parnell 297–8
cod fishing 152–3
Codrington, Professor Kenneth de Burgh 18
Coffey, Aeneas 376
Colarusso, John: *Nart Sagas: Ancient Myths and Legends of the Circassians and Abkhazians* 65
Collins, Jill 348
Collogett Pippin 362
Colnett, John 174
Colonsay 81
Common Ground 336–45, 394

concentrate, fruit 37, 149, 216, 299, 301, 320, 325, 326, 328–31, 333–4, 338
conservation 394–7
Copas, Liz 256, 289, 306–7, 322, 332, 350, 356–9, 395, 396, 407, 410–11, 412
Cornell University 8, 19
Cornille, Amandine 20
Cornish Pomona 57, 356, 360
Cornish, Revd Thomas 294–5
Cornish, Revd Thomas Merton 295
Cornwall 57, 83, 84, 90, *91*, 92, 99, 120, 150, 181, 210, 215, 265, 271, 273, 279, 356, 360, 361, 362, *363*, *364*, 388, *413*
Cotehele House 359–60, 362, *363*
Countryman Cider 381
Courtenay, Sir William 257–8
Coutanche, Julia 285
Coventry, 1st Baron Coventry, Thomas 181
Covid crisis 402–3
Cowley Bridge Crab 257
crab apples (*Malus sylvestris*) 9, 12, 17, 20, 23, 29, 63, 66, 72–3, 80, 81, 82, 83, 119–20, 137, 155, 160, 201, 284–5
crabbing the parson 17, 284–5
Cressyn, Stephen 107
Crimson King 358, 390
Croft, Thomas 150
Crohane, James 144
Cromwell, Thomas 114, 115, 165, 190, 191, 245
Cubbington pear tree 395
Cú Chulainn 77–8
Culpeper, Nicholas 133–4
cultivars of apples 15–16
Cú Roí 77–8
Customs 162 326, 327
Cyder Royal 241–6
Cydonia 36, 46
Cyril of Alexandria 53

DATA – Dorset Apple Tree Analysis 356
dégorgement 226
D-Day 309–10
da Gama, Vasco 140
Dark Ages 88, 96
Darwin, Charles: *The Origin of Species* 293
Davidson, Hilda Ellis 67–70
Deakin, Roger 338
Defoe, Daniel 155, 238, 249–51, 306
Democritus 34–5, 36, 49
Devon vi, 72–3, 90, 104, 107–8, 131, 137, 138, 149, 151, 153, 154, 155, 159, 220, 248, 250, 251, 252, 253, 255, 257, 258, 265, 272–3, 279–83, 284, 286, 287, 288, 297, 298, 300, 314, 315, 320, 340, 341, 358, 359, 362, 363, 369, 378–9, 381–4, 396, *408*
Devonshire colic 279–83
Digby, Sir Kenelm 171–9, *177*, 181, 185
dilution 322–6
Dio, Cassius 81
Diodorus Siculus 83
Diogenes 34
Diphilus 35
dissolution of the monasteries 108, 114–16
distilling 108, 112, 115, 117, 130, 158, 160, 241–6, 372–9, *375*, *377*, *379*
Dix, John 339
Dodoens, Rembert: *Cruydeboeck* 118–19
Domna, Empress Julia 81
Doomsday Book 99
Dorset 8, 104, 105, 107, 120, 131, 151, 168, 271, 273, 287, 288–9, 308, 309–11, 324, 340, 342, 347, 348, 350, 352, 353, 354, 356, 357, 358, 359, 380, 389, 409, 412
Drake, Sir Bernard 152
Drake, Sir Francis 138, 139, 152, 181, 260, 386
Drew, Joanna 341
Durham, Dr Herbert 234
d'Ursus, Guillaume 123
dwarf rootstocks 44–6
Dzangaliev, Aimak 21–2, 23, 25, 26, 27, 30
Dzungarian Alatau 24, 26

East India Company 141, 154
Edmondes, Sir Thomas, Treasurer of the Household 181
Edward IV, King of England 107
Elizabethan Age 85, 117, 131, 134, 136, 137, 138, 140, 169, 179, 224, 386, 395
Elizabeth I, Queen 151, 166, 167, 181
Ellis, Alice B. 236
Ellis Bitter 6, 362
Ely, Jasper 367
Emain Ablach 74
Emerita Augusta, Spain 59

INDEX

Endicott, Revd John 159
Endicott Pear Tree 159
English Civil War (1642-51) 133, 134, 182, 187, 188, 189-90, 192, 198, 208, 217-18, 287, 401
English Paradise 45
Enheduanna 12-13
Eris 40-1
Evans, James 356, 359-60
Evelyn, John 208-9, 232, 282
Pomona vi, 286
Sylva 211
Evernue, Robert de 103
Excalibur 84
excise records 269-73

factory cider 308, 312-16
farm cider 302, 308, 312, 315, 328, 340, 390-1
Felldownhead 361
Feng Li 12
Fernandes, Jyoti 380
Ferrice, Rowland 175
Fillingham Pippin 73
Find & Foster 255-6, 381, 385, 407
First World War (1914-18) 278, 295-6, 308-9
Fitzhugh, Colonel William 160
Forbes, Susanna 379
Forsline, Philip 19-20
Franco-Spanish War (1635-9) 220
Franklin, Benjamin 150, 160-1
Freia *69*
French, John: *The Art of Distillation* 242-3
French, Dr Roger 147, 212
future of cider 401-6

Gregg's Pitt 368, 370, 379, 397
Gresham's College 174, 176, 208, 209, 215
Gaia 40
Galen 32, 37-8, 134
Garden of Eden 11, 13, 25-7, 129, 160, 389
Garden of the Hesperides 40, 41
Gaymers 312-13, 314, 315, 316
Gee, Michael 363; *Devon Orchards* 363
genetics 1-2, 8, 11, 15-20, 22, 23, 25, 29, 235, 292, 293, 364-5, 394
genius locii 57
Geoffrey of Monmouth 84; *Historia Regum Britanniae* (*The History of the Kings of Britain*) 84
George III, King of Great Britain 260, 281
George IV, King of Great Britain 294
Gerald of Wales 102
Gerard, John 119
Gerard, Thomas 131-2, 151
Gibbes, Robert 242, 377
Gibson, Edmund 121
Gilbert, Sir Humphrey 151
Gittisham 131-2, 154
glass making 164-200, *189*
Glastonbury Abbey 84, 104
Glastonbury Festival 86-7, 104
Glastonbury Tor 86-7, 115
Gloucester Orchard Trust 364-5
Gloucestershire 24, 99, 112, 120, 157, 218, 265, 300-1, 315, 353, 364-5, 366, 367, 396
glucose corn syrup 323
Godfrey, Eleanor 175
Godinot, Canon Jean 221-3
Golden Hinde 138
Gouberville, Gilles de 242
Gower, John: *Confessio Amantis* 137
Grammaticus, Saxo 62
Grand Banks 151
Grant, Rose 380-1
Great Storm (1987) 338-9
Greece 5, 6, 13-14, 18, 19, 26, 31-60, 61, 62, 64, 71, 78, 82, 89, 90, 93, 110, 111-12, 201, 213, 242, 305, 349
Grestain Abbey 99
Grossard, Dom 224-5
Groves, Chris 362, 415

Hadrian's Wall 59-60
Haine, Richard 238, 242, 245-6
Halloween 80
Hardangerfjord 71
Harold, King of England 89, 104
Harrison, William: *The Description of England* 120
Harrison, William Henry 161
Harris, Richard 112-13, 114, 366
Harry Master's Jersey (HMJ) 6, 289, 364, 389-90
Hartlib, Samuel 191-2, 195-7, 198-9, 201, 208, 210, 229

Hartshorne, Albert: *Old English Glasses* 167
Harvest cider jug (made by Harry Juniper) *408*
Hastings, Battle of (1066) 89, 99, 104
Hatfield House 127
Hawke, Admiral 148
Hawkins, Sir John 138, 141
Hawkins, Admiral Sir Richard 137-8, 143
Haye Farm, Lerryn 361
Haye, Hugh de la 289
Hayes Farm, Clyst Honiton 71
Hayward Gallery 341
Hayyan, Kabir Jabir ibn 178
Hazard Hill 72
Healey, Denis 278
Heathfield 294-5
Heineken 316
Helen of Sparta 41
Heley, Joe 386-90
Hellens 91-2, 302
Helmsley Market 73
Henleys of Abbotskerswell 314
Henri III, King of France 121
Henrietta Maria, Queen of England 129, 175, 178, 184
Henry I, King of England 104
Henry VIII, King of England 102, 112, 114, 118, 220, 229-30, 366
Hera 40, 41, 57
herbalists 5, 116-20, 128, 133
Herefordshire Pomona 57, 95-6, 236, 298, 302, 374-5
Hermes 41
Herodotus 31, 49
Heron Valley Cider 381
Hewbramble or *Bramble Cyder* 257
Hewes 160
Heywood, Audrey 283
Hills of Staverton 314
Hilton, Polly 255-6, 381-5, *382*
Hippocrates 32-4, 36
Hippomenes 40, 43-4
Hittites 13
Hobbes, Thomas 185
Hockings Green 362
Hockley, Guy 295-6
Hogg, Robert 236
Holme Lacy 173, 181, 182, 183-8, 199, 220, 231, 235, 320, 407
Homer 42
hops 118
Horrell's 320, 381-2
Hôtel St Pol 45
Howell, James 169-72, 181, 183; *Instructions for Forreine Travel* 172
Huguenots 166, 168-9, 178
Huxham Barns 256, 381
Huxham, John 279-81
hydrometer 51, 52-3, 295, 305
Hypatia 52-3

ice cider 71
Iceland 66-7
Ictis 83
Idunna 66-8, 71
Ile-Alatau National Park 24
Immortals, or *Apple Bearers* 49
Inanna 12, 13
Inch's of Winkleigh 314, 315
ingenio 239, *239*
Ireland 5, 61, 62, 66, 67, 74-9, 82, 94, 105, 144, 151, 257-8, 275, 314, 361, 384, 391
Isle of Apples/Fortunate Isle (*Insula Pomorum quae Fortunata uocatur*) 84-7
IUCN (International Union for Conservation of Nature) Red List 29
Ives, Edward 145-6

James I, King of England and Scotland 127, 128, 167, 168, 180, 181
James II, King of England 250
Jamestown, Virginia 156-8
Jaxartes, Battle of (329 BC) 31
Jefferson, Thomas 160, 162-3
Jersey cider 285-90
Jersey Heritage Trust 285
Jocelin, Bishop of Bath 104
Johannsen, Wilhelm 18
John Tradescant & Son 127-8
Jones, Christopher 170
Jones, Dr Schuyler 18
Jorvik 73
Julius Caesar 92
Juniper, Dr Barrie 19-20, 22, 23
Juvenal 55-6

INDEX

Kimmeridge, Dorset 168
King, Angela 301, 337-8, 342, 343
Kingsmill, Richard 158
Kirchhoff, Gottlieb 323
Kivik 71
Knight, Thomas Andrew 165, 232, 235-6, 290-4, *291*, 307; *Pomona Herefordiensis* 293
Kopparberg 71, 405

Laertes 42-3
Lamaesabhal or *Lamas Ubhal* 80
Lambarde, William: *A Perambulation of Kent* 112, 120
lambswool 81, 119
Lancaster, Horace vi, 361, 381
Lancaster, Sir James 141-3, 145
Land of Women 75, 76
Langland, William: *Piers Plowman* 106
Langley, Batty: *POMONA or the Fruit Garden Illustrated* to which was added *A Curious Account of the Most Valuable CYDER-FRUITS of Devonshire* 251-2, 405
Laud, Archbishop William 182
Lawson, John 115
Lawson, William: *A New Orchard and Garden, Or the best way for Planting, Grafting, and to make any pound good for a Rich Orchard; particularly in the North Parts of England* 123-7
La Zouche, Edward 166
Lea, Andrew 22-4, 28
Leach, Beccy 380
Leavitt, David 64
Le Couteur, Revd Francis 288, 289
Ledebour, Carl Friedrich von 16
Leland, John 114
Lenin Academy of Agricultural Sciences, Leningrad 18, 21
Le Paulmier, Julian 121-3; *Le Premier Traité du Sidre* (*The First Treatise on Cider*) 122-3
le Pressurhus 103
Lepsinsk forestry camp 25
L'Estrange, Roger 238-9
Le Testu, Guillaume 139
Lind, James 147-8; *A Treatise on Scurvy* 148
Liscourt, Abraham 168
Little Pomona 379
Lloyd, Frederick J. 302-4, 305
London Horticultural Society 293
Long Ashton Research Station (LARS) 22-3, 26, 302, 304-8, 316, 317, 318-19, 321, 322, 328, 330, 331-3, 333, 334, 335, 351, 356, 358, 364, 371, 373
Longney 365
Louis XIII, King of France 184-5
Lugg, Clifford and Barry *91*, *413*
Lugh 80
Lysenko, Trofim 20-1, 22
Lyte, Henry: *A niewe Herball or Historie of Plantes* 118-19
Lytes Cary, Somerset 118

Mabey, Richard 338
Mael Duin 76-7
Magna Cider Co 104
Magners 105, 316
malolactic fermentation 179, 207
Malory, Sir Thomas: *Le Morte d'Arthur* 85
Malus pumila (small apple) 45
Manners, George 234
Mansell, Sir Robert 169, 170, 172, 174, 175
mantling 196, 203-4, 210, 211, 212, 213, 216, 217, 219, 222, 225, 228, 229, 295
Marcher Apple Network 366
Marmaduke 157-8
Marsden, James 370-2, 379, 397
Marshall, William 139
Martell, Charles 365, 398
Martin, Mary 356, 359-60, *364*
Mary, Queen of England 117
Mary Rose 150
Matthews, Eliza 236
Matthews, Fiona 379
Mavye of Reyle 107
Mayflower 155, 156
Meadyate 256-7
Meare, Mary 247
Medici, Marie de' 175, 184-5
medicine 32-5
Medieval orchards 100-2
Menefie, George 158
meniscus 51, 53
Merlin 78-9
Merret, Dr Christopher 215-16, 217, 227

Mesopotamia 13, 50
méthode ancestral, *pétillant naturel*, or '*pét nat*' 166, 193
méthode champenoise 164, 192–3, 204, 219–23, 225
méthode traditionelle 90
Middle Ages 88
Mill House Cider Museum, Owermoigne 380
Milles, Dean 154
Minchew, Kevin 369, 372
mini Ice Age 165
Mitchell, Peter 365–6
Mitford, Selena 381
Molson Coors 316
monasteries 67, 81, 88, 91, 94, 98–9, 100, 102–9, 110–11, 114–16, 194
  dissolution of 108, 114–16
  monastic cider 102–9
Monkey House 348
Monmouth Rebellion 249–50
Monmouthshire 79, 95, 180, 218
Mont Orgueil Castle 287
Montacute 115–16, 198, 210, 242
Monticello 160
Moore, Arthur 296
Moore, Revd George 145
Morland, George 249
Mount Ararat 13, 36
Mount Ida 41
Mount Vernon 159–60
Much Marcle 91, 301, 302, 313, 370, 379
Mulcaster, Richard 275
Museum of Cider, Hereford 206, 231, 372
Musgrave, Dr William: *De arthritide symptomatica* 279–80

NACM (National Association of Cider Makers) 326, 327, 332
names for cider 6, 93, 240
Nart sagas 64–6, 72
Nash, David 339
National Fruit and Cider Institute 302, 304
National Fruit Collection at Brogdale, Kent 16, 113–14
National Perry Pear Collection 365
National Trust 232, 235, 342, 345, 365
Navigation Acts 165, 259, 316
Nebuchadnezzar 50
Nehalennia 70
Neile, Sir Paul 208, 211, 212–13, 229
Nennius 84
Neoptolemus the Parian: *Dionysiad* 39
Neri, Antonio: *The Art of Glass* 215
Neville-Grenville, Robert 302–3, *303*, 304, 335
Newburgh, John 286
Newt, The 405–6
Newton, Sir Isaac 192
Nibley House 157
Nicholas of Hereford 106
Nineveh 13
Noake, John: *The Rambler in Worcestershire, or, Stray Notes on Churches and Congregations* 284
No Cyder Act Teapot 264, *269*
'No Excise On Cyder' eight day long case clock *278*
Normans 6, 81, 89, 91, 98–9, 103–4, 265, 285, 378
Norse apple legends 61, 62, 66–72
Norton Fitzwarren 294, 296
Norway 66–7, 71
Nowell, Jean 370–1, 391
Nowell, Max 391
Nye, Robert 63

occult 56
Odysseus 42–3, 74, 77
Oidheadh Chlainne Tuireann 78
olive oil extraction 89–91, *90*
Oliver, Tom 368–70
Omrani, Bijan 92–4
Orchard Centre 365
Orchard Link 363, 384
Orchard Portman 121, 132, *133*
Orchards Live 363, 384
Orchard Wyndham 121, 132, *133*
Orestes 53
Oseberg ship 66
Östergötland 66
Ottery St Mary 154
Ovid 57–8
Oxford University 19, 190
oxycrat 32–3

Palmer, Sir Geoffrey 174–5
Palmer, Sir Roger 181

paradeisos 14
Parhams Farm 105
Paris (mythological figure) 41
Parker, Eleanor 284
Parkinson, John 127, 128–31, 138–9
Paradisi in Sole Paradisus Terrestris 129–30
Theatrum Botanicum 130–1
pear cider 71, 117, 402, 405
Pear, Philip 273
pearmain cider 103
Peleus 40
Pendragon 361
Pendragon, Uther 88
Perceval, Edward 174
perry 26, 34, 79, 91–2, 94, 95, 105, 107, 109, 117, 120, 125, 136, 141, 158, 179, 191, 236–7, 259, 261, 262, 291, 297, 301, 307, 315, 320, 350, 365, 366–72, 394–7, 398, 399, 400, 404–5, 406
Perry's Cider 380
Pérignon, Dom Pierre 193–4, 219–25, 227, 228
Persians 13, 26
Petisian 54
Philips, John 231, 238, 246–8, *246*, 280; *Cyder a Poem* 246–8
philosophy, cider 6, 72, 100, 182, 225, 236, 277, 290, 300, 301, 315, 347, 349, 374, 389, 390
Phocylides of Miletus 140
Picus, the King of Latium 57–8
Pilgrim Fathers 155, 156
Pillocks Orchard 100
Pimblett, Elizabeth 231–2, 233, 234, 235–7
Pitt the Elder, William 261, 274, 275, 373
Pitt the Younger, William 277
Plat, Sir Hugh 140–1
Pliny 53–4, 83, 317; *Natural History* 53
Plutarch 36–7, 49–50
Plymouth Sound 108
poisoned apples 61
pomace 4, 52, 205, 286, 301
Pomariums 54–5
Pomme d'Api 54
Pomological Society of France 304
Pomona 56–60, *60*, 317
pomonas (illustrated books on apples) 57, 236
Poole, Nick 349–55, 356, 357, 358–9, 411
Pope, Alexander 252
Porter, Elizabeth 180
ports, cider 105
port taxes 107
Port Wine Treaty/Treaty of Methuen (1703) 259
posca 33
Posidonius 83
Pot Gun cider 166, 212–15
Powerstock Cider Festival 352–5
Prohibition 71, 160, 162, 393
Puabi, Queen 12
pupitres, or riddling racks 226
Puritans 159, 162, 182, 190, 192, 202, 203, 208
Putt, Tom 6, 131–2, 349, 359
Pym, John 259–60
Pynsent, Sir William 274, *274*, 275, 277
Pythagoras 33, 56, 210
Pytheas of Massalia, or Marseilles 82–3

Quicke, Mary 256, 385

Radegund, Queen 94–5
Radio Avalon 86
Raleigh, Sir Walter 151, 156
Raven, Charles E. 118
Ravilious, James 340–1, 343
Red Dragon 141–2
Redstreak 121, 165, 183, 188, 190, 197, 198, 217, 249, 252, 255, 293, 320, 406
remuage, or riddling 226
Renaissance 54, 110, 111, 117, 123
Restoration 208, 260
RHS Rosemoor 363
Ribston Pippin 73, 147
Rich, Jan 380
Rich's Cider 380
Robert, Brother 104
Robert, Duke of Normandy 99
Rodin, Leonid Efimovich 17
Rollo 98–9
Rorabaugh, William J. 162
Rose & Crown near Langport, Somerset 348
Ross-on-Wye Cider 391, 403
Royal Proclamation No. 42 167–8

Royal Society vi, 21, 164, 173, 188, 190, 196, 197, 208–9, 210, 211, 212, 215, 217, 218, 220, 229, 290, 292, 303, 397
Royal Wilding 251–4
Rule of St Benedict 100
Runham Manor 103

Sadler, William 174
Sakharov, Andrei 22
Salem Village 159
Salisbury 147
salt cod 142, 149–52, 154, 156, 287
Sappho 38
Sargood, Adrian 384
SAS (Special Air Service) 297
Saunders, William 281
Save Our Scrumpy ('SOS') 339
Saxons 85, 88, 89, 96, 100, 104
Saye and Sele of Broughton Castle, Lord 242–3
Scotland 45, 62, 66, 80–2, 127, 261, 314, 353, 386, 391, 406
Scottish & Newcastle 316
screw thread 3, 90, *90*, 155
scrumping 11, 40, 41, 147, 313, 339, 385, 409–11
Scrumpy Jack 147, 313
Scudamore flute 170, 182–9, *189*, 233
Scudamore, Lord John 170, 173, 175, 179–90, *180*, 193, 194, 197, 198, 199, 202, 204, 207, 217, 218, 220, 227, 231, 232, 235, 240, 252, 320, 407
scurvy 137–49, 155, 156
Scythians 31, 64
Second Nature 338
Second World War (1939–45) 21, 79, 300, 309–10, 319
Secret Orchard Cider 385–90
Sedgemoor, Battle of (1685) 249–50
Seed Savers of Co. Clare 258
Segure, Michael de 150
Selling Court Farm 229
Severn, HMS 144
Seven Years' War (1756–63) 260–1, 276
Severus, Septimus 81
Shaftesbury Abbey 105
Shakespeare, William 116
Love's Labour's Lost 137
Midsummer Night's Dream 119–20
Romeo and Juliet 136–7
The Tempest 137
Sheppy, Louisa 380
Sheppy's Cider 296, 301, 380, 387, 403
Showerings 308, 315, 328
Shultz, Isy 380
Shutler, Vernon 361, 381
Shutler, Therese 361, 381
sicera 6, 93–4, 240
Sidney, Robert, Earl of Leicester 185
Sievers, Johan 16
Sigfusson, Saemundur: *Poetic Edda* 67
Silk Road 22, 27, 29
Silvanus 57–8
Sink Green Farm 179, 231
Siphnos 35
Skara Brae, Orkney 80
Smith, Ann 365
Smith, Captain John 157
Smyth, Lady Emily 304
Smyth, Richard 156–7
Snow White and the Seven Dwarfs 64
Socrates 14
Sognefjord 71
Solon 46
Somerset *see individual area and place name*
Somerset Cider Brandy Co 41–2, 243, 343, 372–8, *375*, *377*, *378*
Somerset, 4th Earl of Worcester, Edward 181
Song of Solomon 2:5 vi
sparkling cider 4, 164–230, 255–6, 259, 273, 373, 375, *375*, 393, 404, 405
specific gravity (SG) 50–2, 304, 398
Spiller, Joel 277, *278*
Spiritual Use of an Orchard, The 192, 194
Spurway, Revd Edward Bryan Combe 295
Spurway, Edward Popham 295
Stafford, Hugh 251, 252–6,2 57, 407; *A Treatise on Cyder* 253–4
Stalin, Joseph 20–1, 28
Stamp Act, America 269–70, 276
Stanier, Norman and Ann 399, 400
Stanley, Venetia 176
States Act (1673) 287
St Bartholomew's Day massacre (1572) 122

INDEX

St Cadix Priory 361
Steinbeck, John 85
Stempel, Patrizia de Bernardo 70
Stewart, Nigel *340*, 343
St Jerome 93–4
St Konorin 79
St Michael's Mount 83
Stoke Lacy 147, 313, 315
Stoke Orchard 99
Strabo 83–4
Strawberry Norman 147
St Ségolène, Abbess of Troclar 95
Studley, Todd 387, 388
Sturluson, Snorri: *Prose Edda* 67
sugar 24, 37, 50, 51, 104, 119, 134, 164, 166, 186–7, 192–3, 204, 207, 209–11, 212, 213, 214, 215, 216, 223, 224, 226, 241, 282, 287, 320, 323, 326, 329, 369, 398, 407, 411
Sumerians 12
Super Celestial 255
supermarkets 7, 36, 301, 341, 344, 380, 403
Supply, The 156–7
Sussex 99, 103–4, 120, 166, 167, 105–6, 238, 245, 248, 272, 273, 393–4
Sutton Hoo 71
Sweden 20, 62, 66, 71, 405, 406
sweeteners 326–8
Syder Royal 160
Symonds 147, 313, 315
Symons of Totnes 314
Synesius 53

Taittinger, Pierre-Emmanuel 228–9
tannin 4, 7, 9, 20, 22–3, 24–5, 27, 29, 63, 123, 138, 140, 143, 149–50, 201, 205, 254, 257, 286, 304, 316–22, 323, 325, 327–8, 332, 333–4, 357, 362, 367, 369, 371, 386, 390, 392, 393, 407–8, 409–10, 411–12
Taunton 121, 132, 133, *133*, 272, 274, 294, 296, 308, 312, 314, 315, 328, 330, 331
Taunton Cider 133, 296, 308, 312, 314
Tavistock Abbey 108
tax, cider 107, 231–2, 233, 238, 258–69, 270, 273, 274, 275–6, 277–9, 287, 332, 373–4, 392, 402
Taylor, Captain Silas 211, 217–19
Tea Act (1773) 276
Teilo, St 95–6
Tell, William 62
Temperley, Julian 277, 343, 372–8, *377*, *378*, 406
Temperley, Matilda 380
Teynham 113, 230
Thatcher, Margaret 300, 308, 331–2, 333
Thatchers 301, 411
Theophrastus: *Enquiry into Plants* 36
Thermopylae, battle of 49
Thetis 40
Thomas, Alex 24
Thomas, Helen 314, 379
Thoreau, Henry 161; 'Wild Apples' 161–2
Thornbury Castle 112
Tien Shan, Kazakhstan 8–9, 11–30, 31, 85
Timaeus 83
Timberlake, Colin 317
Tirage (addition of sugar and yeast) 224
Toussaint-Samat, Maguelonne 78
Tradescant the Elder, John 127–8
Tradescant the Younger, John 127–8
Tragedy of Bailé 78
Traherne, Thomas 296–7, 298
Treaty of Paris (1763) 260–1
Tree of Life, The 341
Tregarne Farm, Lizard *91*, *92*, *413*
Trojan War 41
Tronchin, Théodore 281
Truck Act (1887) 310
Tupktilla of Nuzi 13
Turing, Alan 63–4, 373
Turnbull, George 196
Turner, J. M. W. 155, 249
Turner, Dr William 118
Tyrrell, Simon 384–5
Tyzack, Paul 168

Ulster Cycle of Irish mythology 77
Ur 12
Urartu 13
US Constitution 276–7

Varro: *De re rustica* 54–5
Vavilov, Nikolai 17–22, 23, 25, 27, 29, 30; *Five Continents* 19
Vertumnus 58–9
Vickery, George 296
Viking 66, 70–1, 73, 86, 89, 96, 98–9

Vilkina saga 62
Villa Las Tiendas 59, *60*
Vincent, Thomas 199
Vinion, John 174
Virgil: *The Aeneid* 74
Virginia Company 157, 158
Viron, Thomas 107
vitamin C 144, 146, 147-9
Voyage of Bran, The 74
Voltaire 252
Vulgate Bible 93-4

Wales 62, 79-80, 94, 95, 96, 175, 270, 272, 273, 314, 353, 391
Walgate, Ben 393-4
Ward, James 249
Ward, Robert 174
Wars of the Roses 110
Washington, George 159-60, 162-3
wassail 80, 82, 119, 134-5, 136, 336, 349, 403
Webber, Jim 308
weddings 46-8
Wellington, Duke of 278
West Milton Mill cider club 348-9
Westminster School 199-200, *200*
West Ogwell's154
Weston, Henry 313
Westons 237, 301, 313-15, 379
Whitbourne, Sir Richard 153, 155
White-Sour 254
Whiteways 315
Whiting, Abbot of Glastonbury, Richard 114-15
Wickham, Teresa 345
wild apple, *Malus sieversii* 16-30
Wilding Cider 380
Wilkins, Roger 390-1
William of Malmesbury 85, 101-2
William of Shoreham 105
William the Conqueror 89, 99, 104
Williams, Ray 307, 322, 332, 356
Winter, Sir John 178
Winter Pearmain 103
Winthrop Fleet 159
Wolseley, Tom 387
Wolsey, Cardinal 112
women cider makers 378-85, *382*
Woodman, Helen 379
Woolhope Club 236, 302
Woolley, Leonard 12
Wootten, William 249
Worcester Pearmain 103
Worcester Priory 115
Worlidge, John 143, 407; *Vinetum Britannicum – A Treatise on Cider* vi, *2*, 226, 237-41, 312, 374, 377
Worthington, John 196
Wycliffe, John 106
Wyllt, Myrddin 79

Xenophon 14
Xerxes, the Persian King of Kings 49

Yarmouth, HMS 145-6
yeast 8, 179, 207, 216, 224, 226, 228, 235, 330-1, 335, 339, 351, 369, 381, 388-9, 390, 393, 395, 411
Yellow Plague or Justinian Plague 95
Ynys Gutrin (Island of Glass) 85
York 73, 115
Yorkshire Greening 73

Zeus 40, 41, 44